MEMOIR IN WATER

SPEAKS THE *WAH* UMKHRAH

THE AUTHOR

Esther Syiem teaches in the Department of English at the North-Eastern Hill University, Shillong. Her publications include two volumes of poetry: *Oral Scriptings* and *of follies and frailties of wit and wisdom*. Her third volume, *Many Sides of Many Stories* will be published shortly. She also writes in Khasi and has published a play entitled *Ka Nam*. Her book *The Oral Discourse in Khasi Folk Narrative* is an important resource for understanding Khasi culture. *Ka Jingïamareh Kob ki Wah*, published by Tulika Publishers is her retelling of a Khasi folktale that she has also translated into English, *The Race of the Rivers*.

She is a member of the PLSI's National Editorial Collective, a grass-root initiative to document the languages of India headed by Prof Ganesh Devy and is the editor of the Meghalaya Volume, *People's Linguistic Survey of India* Vol. 19, *Pt II, The Languages of Meghalaya* which has also been translated into Khasi. She is founder member of the Shillong Forum for English Studies.

MEMOIR IN WATER
SPEAKS THE *WAH* UMKHRAH

ESTHER SYIEM

2017
Regency Publications
A Division of
Astral International Pvt. Ltd.
New Delhi – 110 002

ISBN 9789389605181 (Int. Edition)

Published by : **Regency Publications**
A Division of
Astral International Pvt. Ltd.
– ISO 9001:2015 Certified Company –
4736/23, Ansari Road, Darya Ganj
New Delhi-110 002
Ph. 011-43549197, 23278134
E-mail: info@astralint.com
Website: www.astralint.com

Digitally Printed at : **Replika Press Pvt. Ltd.**

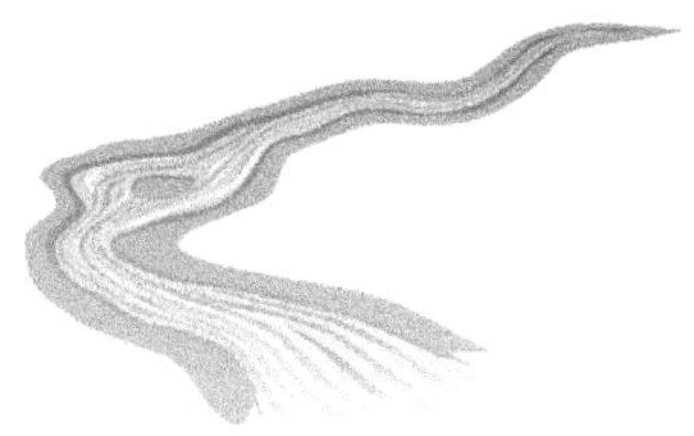

FOREWORD

I am the river Umkhrah, *Ka Wah* Umkhrah, they call me here. I flow from east to west on the northern side of the Shillong valley. My sister *Wah* Umshyrpi in the South flows in the same direction. We meet somewhere beyond Sunapani Falls in the west and join together as *Ka Wah* Ro Ro.

But that's another story. Another time.

For now it's my story

My telling

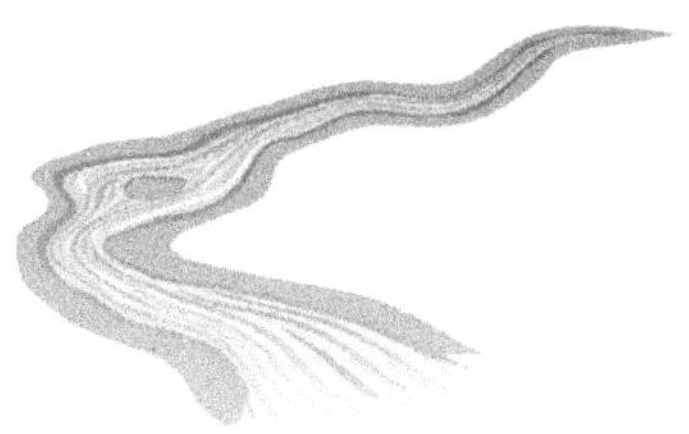

INTRODUCTION

And so the river speaks,
discloses what it sees and hears and feels
to those who think
they see and hear and feel

PROLOGUE

You're the reason for my living, the cause of my anguish, the soul of my future. My life is tied to you as the bamboos that gird our land, the dew that seals our winters in icy wetness.

I've seen what you've never seen, heard what you've never heard, watched you when you were young and innocent; scooped you up when in your despair you flung yourself onto my rocks and battled my currents even as you were drunk with *kyiad*.

I hear you every day and though I'm silted over by your neglect and your refuse and your uncaring habits, I've fought to carry you as I carry myself from one day to another.

Your story is mine, your failures have become mine and your loves have blossomed unfailingly in my heart. That's why I don't drown because this too is the stuff of my life, the reason for my living.

But your illusions? Your illusions, I refuse to acknowledge even as I wash them clean one by one, after you. That's why it's never been easy for me.

That's why I too have to tell my story even if incoherently, even if ineptly, even if, as flawed as you have shown yourselves to be. And though the timbre in my voice has gone, I will outride any disaster, any famine, any plague, any catastrophe to show you that as long as you live, I shall also live.

I am not decrepit.

I must carry on with life and flow onwards, flow outwards, flow to no actual destination, flow as only I can, flow fast, flow slow, flow on to allow your dreams to crest my tide.

I know you've never really needed me. None of you ever had any use for me. It was always I who came to you; I who offered you my soul; I who gave you my being.

My tears sting as I tell you this.

Tears of anger; tears of disappointment; never tears of defeat. I am in with you whether I want to or not, whether you want to or not.

In all honesty I doubt if you'll understand me. These, these stories have to be told. They're not what you'd want to hear, not what you'd expect to hear but what you have to hear.

So don't you dare turn your eyes away or seal your ears from me. My stories shall see the light of your day.

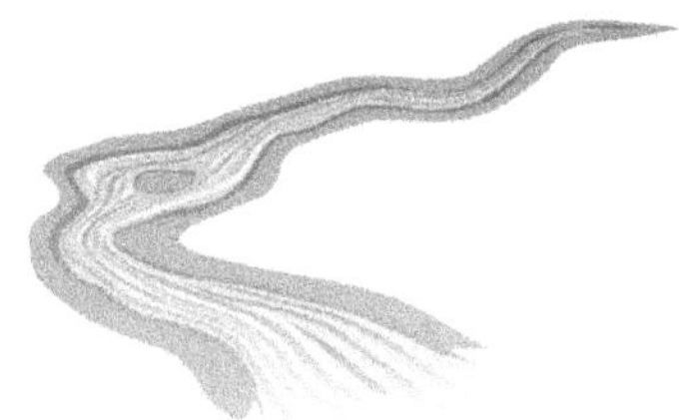

CONTENTS

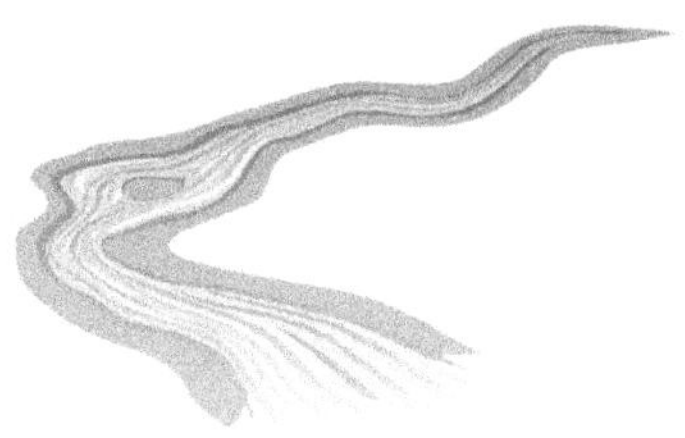

Chapter One

So it was that on that wild raging night, the night of the sorrowful drowning, my banks were feverish with the swollen monsoon. I felt a restless longing for the company of men, thirsty for diversion from the engorged wetness; little knowing what was in store for me, what was in store for all of us, for that matter.

The turning byways and twisting alleyways were deserted and the rain unrelenting. Only the thundering skies and flash of lightning broke the uneasy monotony of the pelting rain, until I heard shouting in the distance, curses and the beginnings of a tuneless song.

In the hazy dampness of the torrential downpour, I saw two friends entwined to each other and making their way towards me. It seemed that they'd braved the wetness on their drunken way home. It was late but they were warmly protected by some recent, delightful memories it seemed that wrapped them up snugly inside; as if mocking the rain, as if mocking even me.

They stopped beside me now, uncertain; uncaring. That was the mood they were in, I tell you!

In their drunken stupor, they sensed that they'd reached familiar ground, so they stood there drunkenly, as if waiting for something, listening for someone. Then they started a mindless tirade that insistently rode the rain and pierced the night and unsettled me.

When I heard their voices pitched against the falling rain, bouncing off me again and again as if they were pitching a ball-game at my expense, I felt a slow anger beginning to rise. I was especially angry with the one whom I'd always loved and protected; the one whom I always thought was sensitive and understood me.

He was totally unlike himself. He was unrecognisable. Or was it that I never really knew him at all? Was it that he hid his real self in front of me and hoped to woo me with his apparent innocence?

As for me, I began to feel the fury rising, rising; a wholly uncharacteristic rage that was difficult to control, that I knew even then had consequences beyond my reach; I felt it.

So I begin a story that I have to tell, a story that won't let go of me until I've told you all. Even when the heavens no longer seem angry, even in the dry coldness of winter when I become a shrunken rivulet, I've never been free of what happened that night.

Of other nights and other such occurrences, they've never disturbed me; never bothered me as this has.

2

So, the rain poured steadily down all through those endless monsoon days and that evening, there was a certain pounding in the rain that was unbearably familiar. It reminded me of those strange nights of untraceable hijackings when unknown forces twisted normal senses and turned them beyond understanding. It reminded me of days when the rain drummed right into our lives and mercilessly conquered everything around, of nights of impounded dreams and days of monotonous wetness that crept into our very hearths.

I remember them only too well!

That evening, as they were hurrying back in the rain after a long day at the office it seemed there was still much to be enjoyed before going back to homes that ticked with reminders and responsibilities; flickering loves and small successes and petty failures.

And in the driving rain, Kongheh's stall was but a natural preference to being stifled at home; especially after a drinking binge at her stall. She served the best *kyiad* ever brewed. She was the woman I never got to meet, never got to see, but I'd heard much about her; rumours about her womanly ways, her charming manners and her cooking skills; enough to make any wife jealous I'm quite sure.

Her stall that night was suffused with the warm soya tang of *tungrymbai* and *dohjem* entrails simmering over a charcoal oven that radiated the laughter of men who've put away their responsibilities for the day. I can almost savour the smacking taste of her cooking when I hear them talk about the food that she served up everyday!

This was where they spent the the best part of the evening nowadays. When they came by my way later that night they were on their drunken way home from her stall.

Somehow when I picture that evening, I'm sure that the smoking scent of a charcoal furnace mingled with the vapours from her brew made it better than home itself; and so Khrawbok and his friend Shanbor spent their evening in the familiar warmth of the cosy stall drinking steadily into the night, amply protected from the rain by the scent of Kongheh's personality.

Kongheh must have been at her best that evening. I knew at least that it was the beginning of the month, the time when I heard you folks talk about nothing else but money earned, money saved and money spent, money talk; as if everything depended upon that thing you call money. So, as I said, money was flowing easily and payments were being made. She could afford to allow her regulars to tank up and pay later.

"This way they can relax while the night is still young, before going back to wives who nag endlessly with no concern at all for their well-being."

I could almost hear the tick-tick-tick-ticking, tick-tick-tick-ticking of her thoughts as she favoured her customers with womanly words dripping with real concern. She had a husband who drank all right, but judging by what I heard, she was just waiting for the right moment to kick him out, the unfeeling wretch.

"What's there to worry about anyway, when I have baby daughters who'll look after me in my old age? I'm not barren that's for sure, and my future's secure. My womb has given me sons and daughters aplenty and my sons too will be well settled before they leave home to marry. Husbands may come and go but children must be birthed for future security. And they can well do without an interfering, irresponsible father."

So, life was good for Kongheh and prosperity was making her careless of a husband who was always sated with drink. He'd been a persuasive lover once, passionately involved in making her life a little easier a little better and as husbands go he'd been good to her. But she began to tire of his growing unreasonableness, his increasing possessiveness.

I'm sure that he too must have loathed her unabated ability to attract other men; wanting her all to himself, not quite sure how things had turned out so badly for him. So they'd drifted apart, selfishly apart.

And as I heard what all those men recalled whenever they came to smoke and rest on my banks, she was one of a kind.

I despised her. I loathed her. I envied her. But what does the opinion of an aged river matter anyway?

Secretly I admired her independent spirit though, but I was wary of these types. She was as ruthless as the rain that night. She knew how to entice her customers who were mostly men, who came with their cronies, who were always ready to be lured by this scintillating image of her standing on the threshold of her stall welcoming the entertainers of that evening; thigh-slapping entertainers, back-slapping joviality that set the tone for the evening; before they all went back to the dull, unexciting routine of spousal familiarity.

Some women, I found in the course of my life, to be colourless and lifeless. No effort to live well. No effort to change the order that their lives had taken. A disgrace to womankind, I'd say! Some simply wilted in the face of challenge.

Kongheh on the other hand, was a fighter, unscrupulous, but with survival instincts to beat any underwater demon! I grudgingly admired her. Perhaps we had something in common?

She stood in direct contrast to those *dkhar* women from the plains who, *bapli,* poor things could never take charge of their own lives. Though they lived in these hills and associated with the women here they never dared stand up to their fathers or brothers or husbands and paid homage even to their shadows. I knew that these women had nothing to boast of, for even the family lineage came from the men. So different indeed from all these hill women!

But then, even here I found inequalities and inequities, differences and disparities that shook their supposedly free world to its very foundations.

But what about the men of her kind? What about Khrawbok then?

"Life's too short anyway to be overly spent with one's wife and children. Friends are the lubricants that set life in motion", he'd always reasoned with himself whenever he was in Kongheh's stall, as he turned that particularly wet night into a drinking binge at Kongheh's in Ïew Polo,

"to quell the tide of monsoon rains in our heads and to shut off the picture" of his waiting children and ever youthful wife. I'd heard about her alright but never seen her yet. She was as you may have guessed the exact opposite of Kongheh.

She'd met Khrawbok when she was still an adolescent and as they say, hadn't even mastered the art of burning *tungtap* chutney; hadn't even learnt to boil rice the right way with each grain bursting with nourishment. It was always *khaw-ot*, as hard as stone. She was also barely out of school and was unfit for any kind of "office" work, never having passed, so I heard them say, the higher classes.

But in those early days of their marriage that was of no consequence, for life was good and, for the young couple food was of no concern, for they were surrounded by unmarried sisters and brothers who gave what they could for the young family. But with the birth of the children, one by one the siblings left to start their own families too.

Things were always tight nowadays and, although she never guessed that Khrawbok frequented another nest, she did notice that he didn't have as much time for her as before. She passed it off, in all her ignorance, as being part of his work; no wait, let me revise what I said just now, she'd learnt to put on ignorance to protect herself. Soo! Even in these parts women pander to men's egos. And she tried to accept it – just like that, for now at least.

She didn't have the womanish charisma of a Kongheh. She was young and immature when she married him and seemed to stop growing altogether in the course of her married life. Khrawbok, had wanted her to remain so, for he took pleasure in having a wife whom he thought he could protect and who would look up to him unquestioningly.

It was a strange relationship, unequal from the very start. It made me uneasy to see them. It made me uneasy to watch them. It made me frightened to see so many such relationships in the course of my searching forays into your human world.

I may be right I may be wrong; I may not seem to understand you – people I mean – too much, but I have an unerring instinct for what's good and what's bad and I've seen that there's plenty missing in your world, so don't you catch me and tell me off ... Do you hear?

She'd given up her life for her husband and the children that she looked after so dutifully, and whom Khrawbok had sworn he would always love and provide for; even though he forgot his promises as soon as they were made, carelessly and without much thought.

Stupid, cruel, unfeeling man. He was so unlike his father. He'd never have done such a thing nor think in such a way! I know. I knew him well!

She knew that he frequented a particular stall after work, but in her, *phuit, phuit*, ignorance, never suspected a rival. Perhaps she was too taken up with the business of raising the children almost single-handedly now; but to be fair to Khrawbok, I knew that he'd been an affectionate father who was wont to give everything to his children in order to make them happy.

Sadly, however, his newly changing lifestyle had begun to alter him in many unacceptable ways, even to me, a mere observer.

Such are men who suddenly lose all sense of who they are, when they are taken up by other women. You'll get to know about it soon don't you worry. He became changed beyond doubt, altered beyond anything, as he got himself deeper and deeper into this other kind of love, what you humans call, *ieit klim*, adulterous love.

And that morning, Khrawbok left home with his wife's bewilderment and unasked questions burning deeply into him. They were reflected in her eyes. She was only just beginning, at this time, to see something she'd never seen before. She'd allowed herself to be too secluded, too cosseted in her husband's familiar love and her instincts too lulled by his over-protectiveness, that it would take a while for her to understand what it was that she saw. But eventually she'd see.

What then? What would happen then?

He'd begun to feel uneasy in her presence I knew, and I'd seen an unfamiliarly cunning desire to get away from her. He was finding her presence more and more difficult to tolerate. But he couldn't leave her yet. Something kept him tied to her.

She'd become a habit, difficult to throw away. Besides he still felt for her in a way that he'd never feel for Kongheh. I saw it in his heart. I knew.

And I saw then that something flickered within his wife. But what was it? I would never know for it soon got over; too soon.

I tell you, I've seen wretchedness in men's hearts, seen some of their miserable ways and wanted to confront them with their own wretchedness. I've wanted to show them just who they are and bring them miserably down but I've always been stopped by my wretched inability to make myself heard.

For who would hear me? Who from you would want to believe that I actually exist, I actually see? I hear?

Wretched Khrawboh left unuttered many words that he wanted to tell his wife. What they were only he knew, but this I saw: he was well able to talk his heart out to Kongheh!

Kongheh had slowly become *his* mentor now and he always felt soothed by her presence. She assuaged his fears with a look here and a touch there. He felt better even if he just sat there watching her flit here and there amongst her regulars. Her stall, at the heart of Ïew Polo, was the centre of his life now. He frequently passed his evenings there helping her out with her customers and I'm quite sure, making believe that he was home at last!

Those were nights when he left very late and I would then see him scurrying back like one pursued. If he so badly wanted to stay, then why return to a wife whom he felt he could no longer live with? I could never understand him at all now.

Only contradictions and denials, contradictions and denials.

And so he continued playing his game, with falsities and concealments, observing meanwhile from his reserved corner in the stall, that all types frequented the shack; the Nepali *daju*, the Khasi *mama*, the Bihari *misteri,* even respectable types like Kpa u Rit and Kpa u Bok who were officers in the big Secretariat office. When they arrived at dusk he knew that they were parched, absolutely thirsty for her "tangy tonic".

He understood so well why the regulars were so regular and his appreciation of her went up a few notches more. She was generous and patient. She wasn't attractive, not at all, but she charmed all her customers with her easy, welcoming manner. Another added attraction for all of them was that she seemed to be at the mercy of a bullying, no good husband.

He was now jobless and careless. He'd left his job for her; he'd wanted to leave his job, to work beside her, work for her. So he'd left a government job to grill and fry and steam and ferment those exotic Khasi dishes that were always pungent and fragrant at the same time and made me want to taste them, even though I can't. Now, it did seem as if he were living off her labours, unlike in the old days when he worked so hard for her.

This made all of them feel especially manly, especially protective towards her. She used it to her full advantage, the clever woman and never lost a moment to show her customers the kind of boor that he was! He'd be so drunk sometimes; he'd be sprawled asleep the whole evening snoring heavily in a corner, next to the charcoal sacks, like a menial.

Only I knew why things had come to such a pass.

However, he was getting on in years. Could he help it if he got addicted to what she'd been selling all this while? Didn't your elders say something about selling *kyiad* –that it would bring both money and curses to the family dealing in it? Blame her or blame him? Blame all of you for that matter. I don't know, but if you were to look at him closely even now, you'd still see dedication, pathetic love for her. All the same he misbehaved and there was a reason to it no doubt. She played her cards shrewdly, exploiting her apparent exploitation to the fullest!

I'm no story teller I know but hear me out won't you? I've seen too much of your world not to want to tell you about it, whether you like it or not, whether you want to or not. And I know that you know but don't want to know, which is why I have to tell. You can never shut me up, never. I have every right to tell and tell I will to all of you!

"All sorrows need to be downed in drink" had become Khrawbok's motto and "especially on an evening like this, all wetness ought to be dried in drink" he told himself.

This had become the very substance of his life and he now moved from one day to another in a hazy determination to keep off the drinking. Sometimes he'd doze off drunkenly on my banks slumped on the grassy slopes, dead to the world around him. There were times when he'd stop by to sit down beside me and mumble drunkenly to himself. He seemed no different from Kongheh's husband then, only younger and so vulnerable!

Somehow he always sought shelter beside me and I tenderly kept him safe, I always kept him safe, you know. I loved him, I've always loved him. There were nights when I shielded him from the cold and the wind with the help of my underwater friends. Then I'd sit up and wait patiently for him to come back to life, my breath hardly blowing upon stilled cataracts. But never did I ever hear words of appreciation from him. Still I loved him for what he was, for what he could have been. He was insensitive to my feelings, however.

I sighed with concern for he was making insane choices that would soon drive him to take desperate measures. But could I tell him anything? Could anyone say anything to him now? I knew, already I knew, but I had to remain silent. And I loved him so much.

He always promised himself that he would come home early and finally shun his friends, but they were too overwhelming in their kindness. He'd surrounded himself with them all his life; divided his life impartially between wife and friends.

They were long time friends, office friends, city friends who believed in living life to the full, habitually concluding their office-days with a few drinks and snacks at Kongheh's. But they knew how to reign themselves in, not to be so addicted to friends or to drink; unlike Khrawbok who could never disappoint his friends, who couldn't resist them especially his closest friend Shanbor, the one he'd grown up with. Strangely enough they were now working together in the same office. Like brothers, they'd been through life's ups and downs together. When Shanbor's wife left him for another, he had no one to turn to but Khrawbok.

What did Khrawbok do but take him to Kongheh's stall?

She comforted him and soothed him and replenished his depleted soul with *kyiad*. That was about the time when Khrawbok took to drinking regularly with Shanbor, only to keep him company, he reasoned to himself at first.

He'd take a regular detour every evening from his office to Kongheh's stall to wait for his friend, be present by his side and comfort him at this critical time in his life. And that was when infatuation for Kongheh began to take control of him. That was the time when a new era began in his life.

I saw it all. And all I could do was fret and whip up my waves in pools of frustration. Did he deserve it? Not at all! But I fretted for him all the same. That was the time when I saw changes I never expected ever to happen. That was the time I learnt about the games that all of you play. Unlike us.

We do nothing to destroy; only when the elements get us, only then are we compelled to destroy and raze down everything you build.

And I began to see that he was drawn to the smoky, grimy stall more than ever before. Nowadays he'd even make time to come for a short while everyday in the mornings, before going to work and he made sure that he ended his day there. He'd just about made up his mind to leave his wife and stay on overnight without further ado, when the monsoon poured its wrath upon the land and left him unable to make the break, for no reason at all if only because of the sheer size of the raindrops.

3

It was as if the rains had put a damper upon his adulterous desires.

A week after the drowning, Khrawbok's wife, Shandora, would shed inconsolable tears on my belly. When she visited his grave nestling high up in *'Lawmali* the pine forest adjacent to me in the slopes beyond, she'd let out her pent-up feelings through tears that fell into the now placid waters below convinced that no one would hear her. But after hearing her tale again and again, I was always comforted by the fact that I'd cushioned his drunken fall with my waves, however swollen, however turbulent and however deadly. But let me tell you very frankly, if that's any consolation at all, that by the time he hit me he was already on his way out.

She sought me several times on her way there, inadvertently drawn to me as her husband had been and revealed her feelings so transparently and so naturally to me. She sensed perhaps that I was a different river altogether?

Never did she ever suspect, never did she ever want to suspect her husband of unfaithfulness, of the treachery that he'd been about to commit. I pieced the incidents together bit by bit, whenever she recounted her tale of sorrow and loss, all to herself as she always thought, and learnt about their brief life together.

It had been full of promise, she sobbed out loudly to no one in particular. I heard her though. Stupid, foolish woman I felt like calling out, if I could. Then I stopped as I understood what she was mourning for; for she knew; she'd known alright but continued to love him all the same. She didn't want to understand how the man whom she thought she knew so well, could have met such an end. She blamed herself bitterly.

What was I to do but listen to her? I knew exactly what knowledge was buried hidden within her and that made it even more painful for me. Damn, damn. Now I was getting dragged into other issues.

I sighed. So many kinds. So many women, so many men. All of them as difficult as the other. And I, where was I?

Always drawn to them always drawn to you with no reason at all.

4

Meanwhile, Khrawbok's life had begun to take a new turn. He was seriously considering leaving his wife.

I knew it just like that.

All this time I'd watched him with eagle-eyed concern and fear and sensed not so much his restlessness as his readiness to spend more and more time at the stall. He himself realised it only after the incident of the lost shawl or more correctly, the lost *tapmoh*. When he found that he'd misplaced it, I knew for sure that he thought he'd left it at the office where he worked in for he searched for it in his almirah there, though of course he never found it. He decided to forget about the bothersome cloth and instead buy his wife a new one. She was elated the foolish woman, or so I thought, as if that was proof of his undying affection for her.

She wrapped it around her at all times and kept it out on the bamboo rack in the kitchen, for immediate use at all times flaunting it as if to tell herself that everything was well. She was trying to wear her husband's love all the time, as if to put it on externally it seemed. She knew but didn't want to know. That was the woman she'd become. The gods take care of her should she be made conscious of it, I remember thinking to myself! This woman was another kind altogether. I found that it was easy to misjudge her because of her comparative youth. And what's more I found her intriguing too. Her love was her strength!

Unexpectedly one night, a long while after he'd forgotten the lost *tapmoh*, when he was at the stall, Kongheh took it out and half held it, half returned it to him as if she didn't want to part with it. He was immediately overcome with surprised confusion and did I sense guilt? I'm quite sure there was plenty of it too but he just wasn't going to admit it to anyone.

I was severely disappointed, though I should have expected it, when he refused to take the *tapmoh* back because as he told her and as he comforted himself, he'd already bought a new one, a replacement for his wife. So the *tapmoh* that rightfully belonged to his wife found a place with Kongheh at her stall. It draped her on certain cold days, as it draped her that very evening–a forked reminder of the hidden byways that he'd now learnt to navigate, the turnings that he'd taken and the switches that he'd made. She wore it every evening now.

That same night he felt the uneasy presence of his wife in the stall. It seemed that she'd accompanied him there, in spirit of course for the *tapmoh* that draped Kongheh, spoke to him in the fissured voice of a wife who had always trusted him even beyond her own understanding. And it hit him as never before. That's when I began to feel something like contempt for him and I watched him even more closely after that. I watched him with a hawk's eye, as you would say.

On that destined night then, when I heard them berating the rain and challenging the heavens drunkenly, I was overtaken by anger, heat-like in intensity that burnt the wetness right through and singed me unexpectedly even in the cold. By the time he and his friend reached me, he'd drunkenly cursed his way out of all acceptable behaviour. They were unusually boisterous that night. This was totally unnatural and uncharacteristic.

The curses that especially he, Khrawbok, belched out seemed to intoxicate him more than ever. Kongheh's brew must have been laced with something deadly that evening, for I could smell its fumes boring right into me.

This was unnatural for Khrawbok, but nothing was natural to him any longer now. Something was upon him that night. His cruelty and drunken callousness infuriated me. This was not his usual self. This was some drunken madness that had befallen him, that recognised no boundaries anymore. Of the two of them, he was the one who seemed to have been hijacked by the storm outside and the storm inside of him.

It was causing him to reel in drunken confusion as it seeped right down to his bones. I saw it. I felt it and I knew that he was too far gone. The thick, blinding rain, the despair, the insane love; all this proved to be the cause of his maddened behaviour. I was irked, driven to the kind of anger that holds no mercy. I wanted to snap him then and there.

Shanbor had no part in all this so I left him alone because I had no bone to pick with him. Khrawbok alone, with his insane behaviour, was my adversary that night.

Didn't he know that I too have feelings just like him? He was brazenly insulting me standing drunkenly there in the rain and spitting out insults at me. I wasn't going to take it lying down the way that unfortunate wife of his always did! I knew that he'd be mine even as he groped his way instinctively to my side in the pelting rain. He lowered himself on the guardrail that protected other creatures from falling into me.

In the thick mist that rose from the furious sprays that lashed me and beat me, I waited for him.

5

The rain fell fast and thick. There was no one, no one around. Even the road-kings of the night, the mongrels that moved in packs and stayed together at night, were silent for they'd found refuge from the deluge. Only Khrawbok and Shanbor were about, daring even the rain, daring all of us, daring me his protector!

I saw him as the indescribable pollutant challenging me in every manner possible. He was challenging the heavens and mocking the powers that be. He'd become a miserable puppet of his own fantasies, deluded into believing that he was stronger than I was, stronger than any of us, than any of the friends that I knew.

I knew that he would fall over backwards into my thirsting waters the moment he tried to get up to rant at me again. The merry-making continued deep into the night and their tuneless singing resounded in the thick rain.

All hell had broken loose within me. That was reflected in the pouring rain. But Khrawbok was too intoxicated to know night from day, me from my other friends, dry land from watery grave. He got up unsteadily to continue his wanton, drunken serenade and climbed onto the guardrail. He chattered insanely, and cursed and sang unnaturally, again and again and yet again; so unlike him and so doubly infuriating to me.

The guardrail was the chosen platform for his performance; the pitiful finale.

This fact of his life, I needed no human interpreter. The moment I caught his battered body I saw what was written in his heart. But he'd kept his passion well concealed from his wife, the crafty fellow.

Later, much much later when my heart was stilled I wondered again to myself. Did he hide it because there was still some lingering, left-over love for her or was he simply biding time till the right moment before he'd break her completely?

I was never to know. More than anything else, I was shaken by the violence of my feelings that night. They broke through every possible deterrent in my destruction of him. To hold Khrawbok lifeless in my arms, to be party to his death!

I was stunned by my own murderous rage, shaken by the violence hidden deep, unable to control my feelings any longer. But how could I, I ask you, control myself when he'd crossed all limits of human decency? He'd shocked me and shamed himself. And yet he still raged on like one insane. He was like one driven!

Later, oh much much later when the damage was done, I wished and wished I could have brought him back for the sake of his wife and children. But I knew that I too was no better than him, no better than others like him. I'd lost control completely and I, like him could not bring anything back. But I did want him back you know because even then I found that I could forgive him. I loved him too much.

The rain permitted him one last-ditch attempt to remain steady, before the wind forcibly pulled him up and dragged him down to smash him violently into my churning waters. Shanbor turned tail and fled. The heavens assisted me in my vengeance and I hid the body deep down in my underwater caverns, my unplumbed *umthlong*, churning powerfully in the replenishing rain. I pummelled it beyond recognition.

Then I waited for the rains to recede before I finally expelled his battered body onto my green, luminous banks now washed by the ripening, summer sun.

"With the ritual of death by drowning the gods are finally appeased and the rains will now abate", I heard you say. "The ultimate peace offering", you insisted.

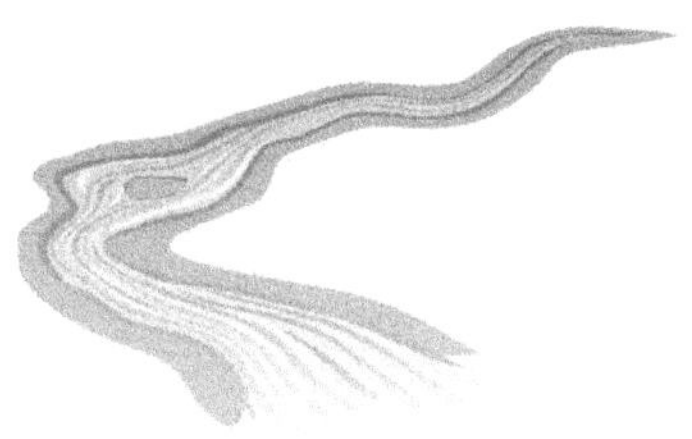

CHAPTER TWO

Merely a river that's outrun its time you say? Does my age show up in my shallow, cataracts? Do I hear you say that they're dying? Have I become a disaster of sorts that you try to put me down like a mad creature? Because I've never refused to carry that which you dump on me so indifferently and smell of leftovers from your lives, I've not outlived my destiny yet. Are you there? So how do I regard you?

For when you say, "she never drowns" you put me back on course.

Let me roll my waves back to probe, probe the past; ease it to stillness since I know you must be listening. And I know that your future is connected to mine. Though you think that you can live, have lived parallel to me, let me now put you on course. I intersect you at every moment and capture your dreams at every juncture. You cannot put me away so easily.

Nothing man-made sealed my embankments then. There was only the wind in the pines, the scent of the woods and the occasional settlement along the way. The whistling *pait pait paitpuraw*, of the *paitpuraw* shook me from my slumber and I awoke with the birds, lazily waiting for the sun to warm up layer upon layer of unopened layers. The tiny *shalynnai* flitted transparently, innocuously in my shallows and the lazy *shersyngkai* dug deeper into me. The sleek, well fed *khasaw* gambolling on the surface brought life to my soul, activity within me like a woman pregnant with life.

As I warmed up in the sun I'd feel blood course through my body and I'd strain to follow the path of the azalea avenue that carried the sound of human voices over to me, strange men and women who'd taken over your land. They spoke of homelands far away, of other rivers and watery expanses across which they came but you were still many steps behind, wrapped in a caul of unknowing innocence.

When it rained I danced destruction upon all of you. I was the unparalleled queen. It was as if my time had come and I could transform anything, anything that I wanted. I deliberately broke my banks and streaked my runaway course with the colours of death.

But I never wished to drown anyone deliberately. If anyone did so, I would say, it was always by unnatural coincidence. Children from Mawlai going to school in Mawkhar avoided me but at Weiïong I'd catch the more daring ones. I loved the children for the patter of their weightless feet as they paddled on me. I played with them and dunked them in shallow waters. They didn't know that I had as much fun with them as they with me, but I prevented them from venturing too far into my underground pools.

In autumn there was sheen in my contours. I was as a languorous snake, sunning myself every morning sated with life's juices. I never preyed on anyone, never, and I was left very much alone. I was given my space as you yours. Then too, I loved those bright, spring mornings when I felt the sinuous bodies of young boys and girls swim across my belly playing watery games with one another. There was innocence in the air but there was also the flutter of heart beats. Milkmen crossed me to deliver their bounty in gleaming canisters. Sometimes I'd trip them up just for the fun of it. I too am almost human you know!

But for now, I wish to remember myself only as the river to whom young women came to scrub their heels on my gleaming, crystal sands. They were like *ka 'tiew lalyngngi pepshad,* the flower that missed the dance. So intent was she upon scrubbing her heels clean with fresh sand on some unknown river that she missed the dance. I fantasise that I was that very same river as young and tender and as amorous as them!

Young men followed these perturbing flowers who were always seeking my banks. They pretended to fish for the trout that played hide and seek in my waters and baited instead, with all the cunning that they had, these lissome belles. I had secret coves tucked away from prying eyes that would never find them and I sheltered my lovers there.

Birds nestled close, for food was abundant. Other creatures too came to savour my shallows and my deeps and I always gave them the respect due to them.

But do you really want to know about it?

After you've been with me through the storm and anger of the experiences that I tell you about, my experiences of all of you, I know you'll want to turn away from me because you'll want to shut me off.

Tell me now do you still believe that you're innocent, that you're without any blame? Tell me again, do you think that by keeping away from me you keep yourselves innocent as the little children that play with me? You're no children, you know that very well. So when will you give me respect as you should? When?

When you were still wrapped in your unknowing you let me be. I loved you for that. Even in death you paid homage to me. You never allowed the spirits of your dead ones to cross me, until a rope was paid out fully, right across to the other bank to enable them to cross me successfully in their journey to the other world.

What's the ritual now? Dead bodies trussed up and dumped without ceremony; newborn babies discarded on my banks; mutilations galore and queer things at night, always at night. My nights are no longer restful as they used to be, but wait you'll hear about them soon enough if you allow me to tell you.

I reminisce like an old river, but why shouldn't I? Do you hear me? Will you help me age gracefully for I can still see those days when your young ones would run down the hilly slopes to wash their produce in my waters before selling them at the Ïewduh Market: cabbages dipped in the flow of a watery life speeding past, radish washed clean, herbs perfumed by the moist fountains of my breath; untainted, reviving, I washed the mud off everything. I was your river. I supported life. I washed all your dirt clean. All I wanted to do was flow with the tide and hurry off or slowly pick my way through sun drenched shallows.

The percussions of a metal plate rolling down the ravines, was the young boy calling his chickens for their feed; what a clamour it was, this constant clucking and squabbling of his feathered clan. On busy days it fused with the canter of the Mawtawar horses just alongside me; beasts who bore their bags of rice to the Ïewduh Market as diligently as they'd done these many years.

They spotted me from a distance. Their neighs always hit the right note to draw me out. The aroma of my body they knew immediately from far away and I always waited for the swish of their tails as at my shallowest points they stepped lightly across. Their masters sometimes sat down to rest, to wash their faces or even to scoop up handfuls of water to moisten their *kwai* stained mouths. This was my daylight world; activity all around. I was the centre of their lives and I thrived on it I loved every minute of it.

And oh! Nepali *dajus* in white long-cloth around their waists used to look like trim, smart girls in white skirts. One evening, a lonesome bachelor trudging wearily home followed this mirage at a respectful distance and admired that she could walk so fast, until he caught up with it and saw the lumps of varicose veins on the mirage's legs! But he did marry the prettiest belle in town and was never so lonely after that, nor did I ever find him pursuing mirages again. I brought them together you know. Matchmaker you call me, do you want me to find you a match too? I've done it for many young hearts you know.

On market-days people swarmed all over me flinging themselves wearily on the thick, grassy green strips bordering my banks; but I always knew when Kpa U Bim, Bah Jrong or Bah Mon were nearby. They had a familiar smell that was partly of the stuff that they carried in their *khoh's* to be sold at Ïewduh. Their pipes hung stolidly from their mouths or lay sooty and grimy in their *pla ïew*. They chewed their *kwai* well and smacked their lips mouth-wateringly. They were good for one another and for me as they stretched out on my banks and talked about life and their dreams. Good dreams won them money, I knew, at the *teem* counter.

Bah Jrong was a bachelor. Was he unnecessarily wary of women? Did he suffer a broken heart? He was almost middle-aged I could see, but he was as strong as any of you young ones now and as ready for life. Could he have taken lessons from anyone? I wondered for a long long time. He dug and ploughed, and carried and strained for the woman he worked for.

She had a grandson who spent hours discussing things, serious things with Bah Jrong, things that educated fools like him missed out completely. Sometimes they'd fall asleep so exhausted, that it would look as if they wouldn't wake up till I stood just a little inside of their dreams to wake them up. That young fellow would never imagine how closely involved I was with him! And yet Bah Jrong didn't really seem to be in need of anything at all. His days were work-filled and his nights dreamless. I knew that for sure.

Kpa U Bim, Bim's pa, was a jolly fellow whose age never showed on his rough, weather-beaten face. He had a wife whom he treated with much care and respect for she shouted at him all the time and threatened to leave him because she accused him of drinking too much. They had eleven children one after the other in twenty long years of marriage. He was hardy and strong and earned good money as a *daju*, a porter hauling sacks of rice and chunks of meat for his clients to and from Ïewduh Market; on a back that was as hard as iron and legs as stocky and strong as a bull's. He secured his moneybag tightly to his belt and he'd often caress it with leathery claws that had nails too long and too curved, just to make sure that it was there. He just wasn't bothered by things like bathing and cleaning that his wife and children insisted upon. Twice a month was good enough for him.

I sometimes saw him on one of his rare days of cleanliness and I tell you he shone like the polished brass in your houses and was as handsome as the handsomest bachelor. He was a jolly one he was; dirty but faithful to his wife and children. He never counted his money in front of his wife though, only in front of me. It was always on my grass-carpeted flanks in the company of the others that he spread out his notes, licked his dirty fingers and counted each note carefully one by one; stacking the ones he'd give his wife to one side. At heart he was as carefree as the waves that played on the surface of my body.

Bah Mon was stunted but he had a heart as big as himself. Though half the size of Bah Jrong he was as agile as a goat. He fretted unceasingly over where he'd get the next day's meal for his children, how he'd feed them, how he'd keep them alive, how he'd find work for himself. He had a disarming smile however, and his lips were perpetually red with *kwai*, his teeth stained with its juice. He was always doing something but I could never for the life of me make out what.

All I noticed about him was how devoted he was to his wife. She was equally small but as fertile as the earth itself. I had a fleeting glimpse of her only once. They had nothing less than thirteen children. That was the reason why nothing seemed to fit him. He'd tie his wife's *tapmoh* around him more for security than for warmth, for it was wound around him even on hot summer days. His clothes were a motley fit for him for he'd probably be wearing his children's clothes. But he kept afloat and I admired him for it. My heart softened for him always for he was truly a fine gentleman.

But no more do I see them now. I wonder where they are. Do they miss me? Do they know that I was always the listening fourth in their group? The one who demanded nothing from them even when they stole my heart away?

And then far, far away in time I recall a group of young men crossing me just there before climbing up to Weiking. Several concrete bridges have been flung across me now, but then, I could be easily crossed on rocks that were strewn sturdily across. They were handsome men and strapping warriors. They must have been riding for some time. I'd lost track of them in the higher reaches of my course but when I caught up with them again, I noticed the tiredness on their faces and sensed a certain tension in all of them. What it was I really didn't know nor did I want to.

They were strangely silent as they approached me, as if they perceived that I knew. I was moved by their insight and I want to tell you right now, that, that's the difference between you people now and those others then. They knew, they always knew. Words were used sparingly then, for they were worth their salt. What do I hear now?

Only babbling nonsense and insensitive words that pierce deep and draw blood causing pain, unnecessary pain.

Someone must have told them about my reviving, refreshing waters for they made straight for me, as children their mothers. So I set out to soothe them well. Till today I've never known the nature of their work, but I was gratified to feel them greedily drinking in what I had to offer. The men replenished their gourds with water distilled by clean sands, and rested as if for an eternity. I wanted to hold them back forever and for a fraction of a moment I imagined I'd call my friends the *puri*, the water nymphs inhabiting my underwater caves, to help me seduce them forever.

But then being me, I knew that would be unfair. I don't play unfair games like you do, you know. So I let them go but not until I'd dallied with them and called upon the wind to disarm them with her caresses. If only I could have held on to that moment for yet another day and another and another, for I was lonely sometimes and longed so much for company of your kind. They left in the late afternoon and I saw them climbing, climbing to the top. Looking up I wished I could have called them back. Memories, what use are memories to me?

So in order to escape these memories I sometimes call my friends from the subterranean world. They live in their cities in the hidden byways of a watery world so different from yours. They know you of course, but prefer to keep to themselves. Sometimes, I see your young men fall in love with those alluring nymphs but these are stories for another time.

Let me feel time flow through me, let me relax while I can and recall only those fragments of my life that I wish to.

Chapter Three

I shall never be rid of them I tell you and as I flow reluctantly back to the past, to retrace my journey, these memories invade my thoughts and disturb me again. But since I've found you to be willing listeners even if not sensitive ones, I shall tell you all before you begin to lose interest in the quibbling fantasies of an ageing river.

You say that my story is only a fable *em*? Well so are your lives, fables of untruth and deceit.

You fabricate lies and deceits and mislead your children into believing them. And then you say that happiness is important! What kind I ask you? Bought happiness; stolen happiness; deceived happiness or what kind might I ask again?

For my story is full of the deceits practiced even by mothers and fathers on their own children. I've seen it all I tell you, all the lies that you use to deceive each other, so that girls like Bem, the one I loved most, or Rympei, or boys like Khrawnam and Bok seemed destined for the slaughter house even while suckling babes.

So, when Bem first came to the *sor*; the town, her accent was as thick as the two plaits that her mother tied for her. And wasn't her nose flat. That's why she was called Bemsynda, Flose Nat: Flat Nose; a nickname that happened upon her when her mother's friends discovered the peculiar absence of a nose of any kind and giggled hysterically over it as they dangled her on their knees and tried to push her soft baby flesh up to shape up a nose of sorts. Don't forget her mother was young then and most of her friends greenhorns like her who were falling over to have children of their own too.

I heard her mother belittle her flat-nosed daughter as they crossed over me when she took her to the *sor* to earn her keep. Nothing mean there just a simple expression of love. She loved her in the only way that women like her could, practically and without any frills.

They expected nothing more nothing less and roughed up their children as if to push home the fact that there was nothing to be expected from life.

So these girls came, as have many girls come and gone, to work as maids in the city households. A few I knew intimately, the rest I heard from their friends and family.

Their lives were tied to each other in peculiar, coincidental ways; though I myself hate that word because it passes off too easily so many wonderful, fascinating things that you've refused to see.

That's why I find you so dull, so unimaginative, so unable to cross over to my world. D'you think that life begins and ends only within the aridity of your hearts and the wellness of your lives on dry land? Can't you understand anything better or see anyone else living in another world, someone like me who watches over you? For these very same girls may have crossed your paths in the past or may even become part of your lives some day.

These girls led lonely lives, confused sometimes, but never, never uneventful, riddled as they were by unforeseen mishaps and incidents. They were waylaid by them, pressurised by them. Yet they recovered fast and were on their feet again before I could so much as stretch myself up to my full length.

And the wonder was that they could still love so much.

"Why do I get involved? Why do I take such interest in them?" I've asked this of myself several times over.

I've received no answers. I give my heart to each one of them. And every time, every time, it comes back battered and bruised. Unlike them I take time to recover. But have I ever learnt? Have I ever learnt? And when I smell sickness and a certain desperation fouling their breaths, I follow them slavishly and uselessly, like an animal, a dog, panting after them; as if that would in any way help them at all. I feel that I haven't done enough, haven't cared enough.

They were hard-hearted in their ignorance, mean sometimes, but I knew, I knew that they never meant any real harm. They only succeeded in burning themselves, singeing their lives over and over again. That's why I've stayed with them for I can't be angry for long. These were children who'd been left behind, orphans by default, left on their own, left to find their own way in the cunning mazes of the *sor*. And no matter what as I keep saying, I can't be angry with them for too long!

You say I've become soft in the head? Well what about it? So I have, what do I care? What do you care for that matter?

Her mother was upon her third or fourth husband, it doesn't really matter does it? She had to find someone who could provide for her and the other small ones; and besides, Bem was old enough to fend for herself. These were people whom I often heard you *sor* folks call *nongkyndong*, villagers with no fancy manners. But I knew them all well, knew why women like Bem's mother, had to do the things that they did, like finding husbands and sending off their children to different houses in the *sor*.

Children when they came into this world, so it was said, brought their own plate of rice with them, so there was no worry there. But the hope that flickered unabated in the hearts of these women was that, one day would come a man who would change everything around. So they had to keep trying, foolish ones, and hoping and in the process had more children than they could handle. Not that they didn't love them. They loved each one of them; but sometimes there were too many, it was difficult to keep track.

As the siblings multiplied in number, the older ones learnt to take on the younger ones and these younger ones still younger ones. And the same values got passed on so that these same children went looking for love as early as possible; never knowing that they'd get trapped in the same cycle as their mothers and fathers who were caught by love too early in their lives.

And so Bem had to be settled soon. There was a family that needed someone, a "domestic help" they called them, to look after the house and perform innumerable chores. She was taken away quickly. I heard her mother tell herself that "she'll be a good source of income for us", priding herself in being able to secure employment for her eldest, on the day that they crossed over me to go to the *sor*. After all that was all there was to their lives wasn't it?

The moment they could walk they were ready to start work. As strange looking as she was flat nosed, she was as diligent as the day and sensitive even to the pets she was made to look after. They were funny looking puppies with bones made soft by disease, their fore-limbs curved in a convex. They suffered from some kind of deficiency and she had to care for them night and day feeding them, cleaning up after them and bathing them as if they were small babies. They recognised her voice and nipped her ankles playfully whenever she fed them, for they knew her as a kindred spirit; this girl who'd been an accident in her mother's young life and yet whose young heart seemed so sturdy, that it made me afraid and uncertain.

"No work is too demeaning for her", said her mother to the *sor* family who was impatiently waiting to put her to work immediately.

Her mother left her that morning; happy and relieved that she'd put her daughter on the right path to life. Bem, meanwhile who had had no say in the matter was left like an orphan with no one at all to confide into.

Her only defense against the world when it got too much for her was to sulk and sulk and sulk; for she was not given to speaking too much, except on those rare occasions when she told her stories in thick accents to her middle-aged mistress with whom she had a strange affinity.

They were sad and funny stories of drunken step-fathers and empty stomachs, of nights of shameful sights and overcrowded rooms; days of petty thefts and happiness with friends who shared their lives with each other. Her mistress was the only one who could make any sense of it all. She'd been there and seen it all. And Bem, like all the others never really saw the other side; the privileged side I mean that disgusts me no end. She accepted it all with the kind of good humour that made her strangely resilient, strangely alive.

The children preferred to leave her alone. They were too superior, too protected, too full of their *sor* ways to pay heed to this "*nongkyndong*" as they called her. But I, like the mistress, was both spell-bound and enthralled by a certain spark in her. Her stories were my lifeline to your world. No wonder that I was so drawn to her as I'd never been to anyone else before.

These were times when I felt like a mother to her, fiercely protective.

And so Bem, diligent Bem, vagabond Bem, sensitive Bem was also made to sweep, mop, and scrub and severely disciplined and told to keep her *nongkyndong* ways in check. But did she understand what that meant? I'm sure not! How could she when she didn't know how. Her whole day consisted of scrubbing and cleaning and carrying, scrubbing and cleaning and carrying, till her hands were raw.

She'd never known anything better anyway. She gorged herself on good food from their lavish kitchen. She wore clothes that were handed down from her mistress's children. She was expected to pick up the civilised ways of servants serving their *sor* employers. Her mind was awhirl with new surprises but she allowed herself to be kept busy scouring the same pots, wiping the same floor, cleaning and washing the same endless clothes.

Of course it wasn't a bad life at all! She'd ape the mistress's children in public and try her best to behave like them. But she didn't see that she was separated from them by an unbridgeable gap. They knew they were better and made it obvious to her. But she wasn't one to understand such things. She'd been neglected, left alone to find happiness by herself. She'd been imprinted by her mother's ways. And sure enough she moved with instinctive certainty towards the things that her own mother held important.

But she'd skipped childhood and never gone to school. She never knew what she'd missed; there was a breach in her life that would never be filled. She never knew it. She was also precociously mature for a girl her age for her mother, as I've already said, didn't see the need to entertain any of her childhood whims. She herself needed Bem to grow up soon so that she could be put to use, just as she was at Bem's age.

Bem knew more about the underside of life than any of her mistress's children. She'd been made to work from the moment she could. There was nothing to strive for and all she knew of happiness was the kind that her mother went for!

Having known nothing else this was what she set out to do. This was the path that she recognised only too well and like those puppies she would sniff the air around her to find that freedom for herself.

So at the age of fourteen would you wonder why she fell madly in love with the first man who looked at her? Oh keep quiet and keep your morals to yourselves you uptight hypocrites. I knew these girls as well as each intricate turn of each of one of my waves and I knew the hunger that ate into their lives so that they were forced to come to the *sor* to serve all you rich, fortunate beings.

They were like bonded creatures, as ignorant of life as the little birds that preyed for food in the shallows in my sides; bonded to nothing less than the work that they had to keep doing, keep performing, keep repeating at a never ending pace. But she was as genuine as she was untaught and a good-hearted person altogether. She was a strong worker but she of all of them, would never learn the tricks of the *sor* as did many of the other girls who took to its ways very quickly indeed. Strange to say, she was a dreamer too, an idealist; not a chip off the old block at all.

Wait and see and then you'll understand.

This was why I had to keep her in sight.

She was unusual, don't you see? I knew it from the moment I saw her and sensed that she had this unusual side to her which was un-like anything I ever saw. She wasn't like any of the girls who were able to manage their lives quite well and actually made good marriages sometimes. She was rendered impractical by an unusual ability, how should I put it to you or should I say that she had a sense of things to come? But was it to her advantage, this unusual sight or whatever it was? I asked myself knowing that even I couldn't come up with any answer. And who would tell her about it? I couldn't tell her. I was only a river, only a river that washes down hidden vices and inadvertent failures.

There was definitely something unusual about her though, that I still couldn't quite place. I was drawn by the aura surrounding her. I could almost touch it; and I felt the eddying currents of my body lapping up to it. But she poor, ignorant, untaught child of hardship and labour never knew what she had; never knew whether life would be a blessing or a curse for her. So she would sulk when the knowing became too much for her. She didn't understand it at all. Then it was impossible to get through to her. On such occasions, I'd hear the storm and fury that she caused but at such times she was hard as a *kwai* nut, closed to everyone in this world.

These girls gave me a lease on life. I was one with them. They never caused any conflict in my heart like some of the brash, callous young men and *sor* girls who crossed me and challenged me and hurled blasphemies unashamedly against me and my kin. I cared for them, motherless as they were and too easily influenced by the *sor* ingratitude that infected some of them too, too soon.

I saw them in my sleep, heard them when I was awake and tried to keep them from desperate futures, the ones that they tried to create for their own selves. But I more often failed than succeeded. I tried, however, for my feelings run deep you know and I must say that my insights are sharper than any of yours. But as inevitable as is the flow of my currents, so were their lives as foreseeable as the stars that looked upon me on clear winter nights. I would shiver as only a river can. I was always affected by what I knew would happen; and these girls, these girls would never learn! None of them ever knew, none saw until it was too late.

So she daydreamed too about the unreachable things in her life like marriage and happiness, *dharas* and gold lockets like the ones her mistress's children wore, a husband and a houseful of children. These daydreams made life more bearable for her.

She'd grown quickly with all the good *sor* food and wasn't she strong! I'd hear her beat the carpet free of dirt every week and see the dust rise up in the air; she enjoyed scrubbing the concrete compound with lime powder and water until it was bleached white. She'd accompany her mistress to Ïewduh market every Saturday to carry vegetables and other groceries in cloth bags especially stitched for such marketing purposes.

Those were the days when there was not a single plastic bag waiting to strangle life; strewn all over my body and choking the free flow of my whole being. Those were the days when she met her butcher at Ïewduh.

He would sit like a king in his stall, a seasoned lover with current and past wives tucked away in all corners of the town. His biceps bulged out vulgarly when he hacked the meat to pieces or adroitly carved out different cuts for his customers. He was gregarious and loved surrounding himself with Ïewduh cronies and hangers on. And he knew just how to push a bargain surreptitiously with his customers. Oh they loved him for it. His oiled hair was piled up high in front like a small mound that was slicked up with some cheap, perfumed oil. He was her instant dream man.

On market days when she accompanied her mistress to buy meat and vegetables no words were exchanged. He only gave her furtive, significant looks that said it all to her; wooing her in the best Ïewduh fashion; but her mistress was too involved in the buying and haggling to be aware of what was going on.

He was an adept performer, a Ïewduh performer who performed for customers and would be mistresses and friends alike. He had a sure sense of conquest over his customers and friends and cronies, all and sundry. Basically, I saw that he had the key to his own happiness, however unscrupulous, however mean, depending upon how you look at it. But wasn't he prone to the bottle!

She would be his next conquest I knew. I sensed it then and I waited with bated breath to see what would happen, knowing what would happen, hoping it wouldn't happen, anticipating what would happen, wishing that I could prevent it from happening, that I could seize her then and there, and hold her to myself, hide her from this preying butcher. But he met her every week and made the most of it. He passed on delightful little presents that didn't cost him much and gave her every reason to believe that she was everything to him.

Before I knew it one year had passed. I saw them getting desperate. Since both could never get the opportunity to speak to each other for long, I felt the desperation in them. Not a person knew about it. Her mistress's foolish children never imagined that Bem could fall in love. Oh they believed that she was capable of running amok if unsupervised but they cruelly and deliberately closed their eyes to her needs. They didn't want to know how starved she was of natural affection; that which was the breath of life for them, that which they all took so completely for granted.

And every week it was being offered for free, like those stolen pleasures, those plates of rice that her kindly neighbours offered her when she was a child. I wanted to snatch her away from him, wanted so much for her to see what I saw, but I couldn't, I just couldn't. I was hindered because I am what I am, don't you see? Do you hear me at all?

One night I heard screaming and shouting as if the world had gone mad when it suddenly discovers something unpleasant, something it doesn't want to accept. I quivered unnaturally in the dark as if the wind had raked my waters dry and left me naked and vulnerable. Mean things were said by the mistress's children of all the others, about Bem, disturbing things that sent ripples and tore me apart. I heard Bem's name being spat out in the dark almost as if she were being violently beaten. And I put two and two together and knew instantly.

I was enraged. Who did they think they were to deny Bem the opportunity to live? Was she happy? Did they ever care to ask? I felt my hackles rise but I was powerless as Bem herself; condemned only to watching, condemned to flow with the tide. I smashed myself hard against the banks. A part of it caved in and passers-by commented on the violence hidden in my waters.

What violence wouldn't I have done to such callousness as theirs? What wouldn't I have told them had I been given a tongue such as theirs! But I knew also that I had to reign in my anger, reign in my impatience with these girls who were so blind, so obstinate.

I didn't exactly know what had happened. I had to wait and watch even though I knew something had happened.

2

For a long while I never heard of her again. Why was it I hear you ask? That was because I got caught up in the life of another girl who came all the way from the Bhoi lowlands, the rice bowl of the hills that drained themselves into me. These girls chose to sit beside me for hours together filling in the little details of their daily lives with each other. Whenever they came, I stilled my flowing currents and quieted them for some time, as I gave these girls this space, this quiet, this time, this giggling exchange they were otherwise never permitted to have.

So I learnt about Lili and the household that she went to.

She was slow-witted and dull. Her older sisters were married and the only younger sister that she had was "smart enough not only to catch any man she wants but can do quick calculations to sell her produce at a profit. She's even bought more rice-fields for herself! She can manage very well on her own".

Lili was the stupid one. So was the youngest who was a boy.

"She neither knows what the word future means", said her old widower father, strained beyond his capacities with the responsibility of looking after three young ones after the death of his wife, "nor can she add two numbers together. How will she survive when I'm gone? She who's never had a mother to teach her anything at all; I despair of her! "

So the *sor* was the only destination for her. "These *sor* folks" he'd been carefully informed, "can hardly carry a teaspoon of water from the tap, let alone wash their own *jaiñkyrshah*. They're always on the lookout for strong village belles to do their dirty work and to keep house for them when they go to office. They also pay well".

What a situation. It made me smirk to think that there were people who could be so dependent on others who were not even their own kin. And to pay cash to get care and attention for their own selves! And then to make so much of kin and clan and family; what a laugh!

And so one day, Lili found herself trudging by her father's side, all the way to Nongpoh and then catching the *Bajaar Bos* to Shillong. She was brought to a family that was living close to me. Their backyard overlooked a shallow stretch of my waters and I would be awakened by the daily tinkle of cups of tea being stirred, clothes being scrubbed and beaten, sometimes varied by melodious humming and the sound of boisterous, noisy laughter.

"Stay well" Lili's already bent old father told her encouragingly, "and serve them well. These are good people I've brought you to and before you know it you'd have learnt everything that you need to know and you'll be able to stand on your own."

He'd philosophise with the mistress whenever he visited his daughter. The master was never at home. One day, however, when he was at home, he was given a packet of expensive cigarettes all for himself. "Ah! What delicious change from my *biris*" inhaled the old man appreciatively.

As he dragged on the cigarette and puffed out the smoke, he warmed up to his pet subject and congratulated himself again in front of them: he'd been both mother and father to his children when his wife died at childbirth, after the birth of their youngest son. He was worn down no doubt, but he'd also had a good life. He'd toyed with the idea of taking another woman but as he told himself shrewdly, the sun was setting for him and this was no time to be washing diapers and baby blankets. They'd never dry in the tired rays of his setting sun.

So he stayed with his children and tried to teach them how to survive. He missed his wife, missed her company, missed the *ja* that she would have waiting for him when he came back from the fields. He wasn't a sentimental fool like me. Yet I liked him. He smelled of freshly turned loam and his feet knew the ground upon which they walked; bare feet that were strong and earthy and pitted like the earth itself. His older children were all married and he didn't expect anything at all from them. He knew it would be unfair.

No matter what he'd been before, something there was about him now that accepted things unacceptable before. He requested the mistress to watch over Lili. She gave him her word that she would look after her like one of her own. He was well onto his sixties and still had responsibilities over his other two younger ones, "unfinished responsibilities", he told her, "that won't allow me go off so easily, to join my wife". And with that he left his daughter in the care of her mistress.

He was a farmer with not much time left no doubt, but did he admit it? Never! His feet were as bare and as calloused and cracked as the earth that he prepared every season for the sowing. I felt as if I'd known him all my life and conversed with him on matters close to our hearts. I agreed with him on many things.

At a time in his life when he should have been surrounded by grandchildren, he was still scratching the earth for daily subsistence. These were the children of his old age and he had to make a bad situation work. He'd decided to do it alone anyhow and if he grumbled, it was all part of the game. So a bit of *kyiad* now and then, to keep him warm in his old age, a bit of celebration to remind him of his young days and to fortify him for whatever work awaited him was also part of the game. Had his wife been alive she would have shouldered his responsibilities with him. But he was alone and he had to think about his children's future, though he couldn't help cursing the youngest son now and again for he was a bit of a fool.

So she stayed and worked for them; another *nongkyndong* girl who had to give up the best part of her life in order to serve others. They were good people too despite being *sor* folks; at least they kept their word unlike other *nongsor*. But, try as they did, she couldn't help it, her eyes slowly opened to what she saw around her and she imbibed many things just by talking to other girls here and there.

I'd hear her giggling unnaturally with the other girls and watch the change take place within her. It hurt me as it would her father but like him could I have been able to change anything? I was tied to myself imprisoned within a body that forced me to follow the course I had struck out a long time ago. I was stuck to following this path, forced to flow on and on with ears that heard and eyes that saw everything that even the darkness couldn't hide.

Several times I itched to tell her many truths. But would she want to hear me out? Would she even have the same consideration that she had for me as when she first came? She'd put me out of her mind altogether. I was out of her sight too. She no longer had any regard for me. I was now simply the river flowing past the house that she worked in. Just that!

I understood her father's concern only too well. She'd crossed over so completely to *sor* ways that she neither heard nor saw what were honest reminders of her past. So she lapsed into a sort of indifference that was tough to break, that formed a thin sheen over her ignorance that brought about its own misery the foolish girl.

When her father died many years later, I heard them say that he was most worried about this foolish girl of his. He didn't want to leave her for he'd already read the signs to come when he last visited her. She'd curled her hair and bobbed it in the *sor* fashion; painted her nails scarlet and was wearing some new fangled skirt that was the current fashion she told him. To add to his anxiety she was talking about taking up lodgings elsewhere for she wanted more money working on an hourly basis, a *nongtrei kynta*.

The now wizened, wheezing old man who knew the extent of his own daughter's foolishness made an effort to visit her one more time even though it nearly killed him, made her master and mistress promise him that they would never let her go.

But can anyone be forced to do something they don't want to? Can you tie up a dog that has always been free? Can you dam up my waters and expect me to be still happy?

I've seen how all of you balk at having to do something that you don't want to. Lili was no different from any of you. She was wise to the ways of the *nongsor* now. It was just a matter of time before she too would show the *sor* that she could stand on her own; her shoulders were strong enough to bear the responsibilities of an independent life! Why did it always happen so? Why? Why? Why?

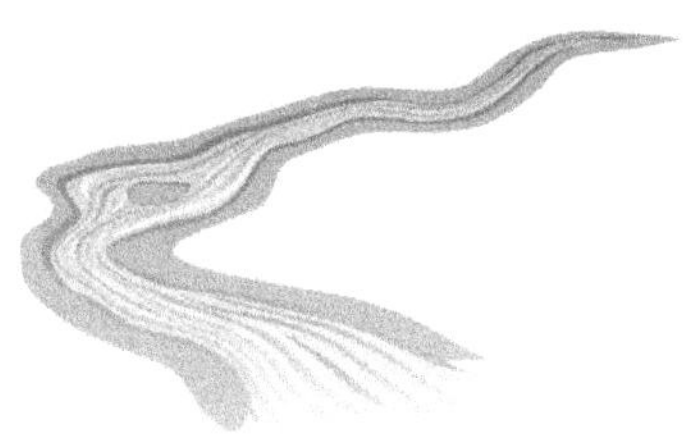

Chapter Four

In those days dusk would come with the yap yap of dogs and the howling of foxes from nearby 'Law Mali. I'd hear mothers shout and call for their children to come home quickly, for fear of the prowling shadows that lie in wait for two-legged creatures like them.

I heard talk of U Ksuid Tynjang, the unlucky demon – unlucky for those who meet him – who catches people in order to make them scratch his back all night, all night long, till blood spurts freely out from torn fingernails of scratching hands. Nights are pitch dark and always cold. But I've never really feared the dark like you do. I accept it as part of everything around me, a necessary part.

Do you? Or do you run away from the darkness to get lost in another darkness that sends you into the kind of madness that destroys?

Because that's what I see day in and day out, madness the kind of madness that I can't understand for sure that seems to spring upon you like a disease or lies in wait for you forcing you to do things that you don't understand. And when I speak of madness I'm wont to tell you of another tale that cries out (to me at least) with confusion and sorrow and misunderstandings and what have you!

This is a story of madness, not the madness of love as you would think, but the madness of pride, the madness of mothers and sons turned against each other.

And who, I ask you is the person we must be sorry for in this case, the mother? She was uninspiring, beaten down, allowing circumstances, allowing the past to defeat her; or the equally stubborn son who loved his water nymph, his *puri* wife to distraction.

I'm not one to judge. Nor will I ask you what you think. But only see for yourselves as you follow me into a world hidden in the watery shadows of a reflected sun. You don't deserve to see any of it of course, yet it concerns you as do many things that you don't care about.

This I've learnt to accept.

2

I wish he'd been my son. I'd have welcomed his wife with open arms. What matter that she were not of this world, he loved her didn't he and she him? They had children together and, despite the hurdles that came his way, he nevertheless, wanted most of all, for all of them to accept all his entire family.

I know, I know, I'm moving ahead of my story; so let me stop for a moment and settle myself comfortably down before I give myself up to a tale that still makes me weep.

He was drawn to the water. He loved to go fishing. He especially loved the magical swirl of the currents in the monsoon rains. They held him spellbound, stilled him, pulled him to their depths and transformed his world so much, that he forgot himself completely.

That's how I got to know him, got to see him too. I loved him as any mother would but I always felt that I never really got close to him. There was always something about him that was as impassable as the ravines of Rngaiñ.

There was certain stubbornness about him, a certain silence that overrode steep cliffs and deeply descending ravines. There was a *jingdukhi*, an obstinacy alright even at this young age that was difficult to break but I tell you I felt I knew that boy, and I knew that deep down he had a sensitivity than many didn't have. It was all so complex and heartbreaking for me.

He was different those days, I know, less affected by life for I'd find him hunched up for hours waiting for the fish to bite, in no conflict with time or anything or even anyone for that matter, whilst seeming to concentrate very hard on the task at hand. It didn't matter that sometimes the fish never bit.

What mattered was that he seemed to be naturally alive and responding to the elements that whirled around him and enclosed him in their tight embrace; even then a natural part of the quiet and peace that surrounded him; his spirit released as it were.

I loved to imagine that he came looking for me, looking for what I knew could never be found in his own world. I recognised his footsteps. They were firm and strong and young and deliberate, but boyishly impatient still. And I found that he was always alone.

And wasn't I concerned about that?

Why didn't he have friends? Why did he seem to be so inward looking? Didn't he have any girl friends? For a boy his age, he was too serious. I thought I'd seen them all, their moods, their loneliness and their ups and downs. But I was wrong. Kyrmen Skhem drew a tight line between who he was and what he wanted the world to see of him. I saw him seal up his feelings privately inside. Perhaps he needed the company of boys his age?

His mother was tied to her weaknesses, *ani ni ni* foolish woman. I didn't know whether I liked her, but I pitied her alright. She'd never known what simple happiness was. Much as I, innumerable times, wanted to drown the woman, drown her violently indeed for all her foolishness and stupidness, the picture of her as a child would always float up to the surface.

She had been bright and intelligent, born in the midst of five boys, vivacious and alive but whose vivaciousness and aliveness got crushed under their father's rule of thunder and lightning. He possessed a voice and personality that everyone said was like dry thunder, *"u pyrthat rkhiang"*. I heard him being described so by everyone.

All I need tell you is that their mother died early and she and her brothers had had to live out an existence full of the pitfalls of a possessive father and a craftily-creative, step-mother. *Phuit!*

For being unable to break her father's hold on her Kyrmen's mother lost out on her children completely. She withdrew into herself so often that it was as if she were given to nasty spells of crazy moodiness. Truthfully speaking, she was so unable to stand up to her father, so used was she to treading on the edge that this had imbalanced her somewhat. Even though there was a break for her when she married her husband that was too brief an affair.

She was unsure about everything, prone to an unnatural kind of introversion that dogged her. The past kept coming back, to flow right back into the present. Come to think of it this unnatural, brooding propensity distilled itself into Kyrmen and handicapped him in so many ways; as it turned into sullenness of a perverse kind.

Now wait, let me get myself clear. Kyrmen had lost a father who could have been his guide, his mentor, his friend, a parent after all who'd counter his mother's influence and set many things right. But as things were, that was never possible. And as things went Kyrmen found himself having to deal with a mother who was so conflicting that by the time he started reasoning things out for himself as a young and growing boy, he found that he was often puzzled and hurt.

Such were life's ironies; for, in the ultimate reckoning Kyrmen took more qualities from his mother than from anyone else, as did his mother from her father! And there were times when he felt older than her and more responsible. It was a terrible mix-up of sorts. And somehow he understood even then but never spoke about it; he just couldn't.

His old grand-uncle, who was his grandmother's younger brother, had recently come to live with them for there was no-one else to look after him. Never having had a son of his own this old man had a strange, special affection for Kyrmen. There was a certain quality about Kyrmen, I heard people say time and again, that reminded them of the old man himself. But these were stray remarks from bystanders here and there who only saw what others saw, not what I saw, not what I knew; about the old man I mean who wasn't at all like Kyrmen.

They were worlds apart. And the family circumstances that brought them together for a time did nothing to at all to shape Kyrmen in any way. But did the world understand this? Never I tell you.

Of his father, only this I want to tell you. He came like a whirlwind and left as he came. A few short years of happiness; only long enough to see his two children together before he died. And then his mother was alone again.

And so what could she do, but look to her father again and because she was after all, his daughter, she slipped into the old habits once again. Her husband's influence had been too short-lived. It was easier to step back into her father's shadow.

She didn't want to know anything better, I suppose. By the time the children had grown up she was set in her ways. She refused all other marriage offers. Her heart had closed up and she'd locked in the happiness that could have blossomed with the children.

Anyway who am I to say such things? All I want to do is to tell you all about my Kyrmen.

And so it was that it was his mother who was the inadvertent cause of much of the suffering in the family. But even as I say this, I know that that too rings false. The circumstances of their lives were to be partly blamed. But Kyrmen too was partly to blame for he couldn't help it if some of his own mother's qualities rubbed off onto him and made him do the things that he did.

They came from the same mould after all, the whole family, all of them.

There was a gap in her communication with Kyrmen as well as with her daughter for she'd fumbled in her responsibilities as a grown woman. From father to husband she'd gone and when he died she had nothing within herself to come back to. So the children had to make do with growing up under the influence of a grand-father who was still unrelenting as dry thunder even when he was past the age of being able to do anything much. Mercifully he died when they were still young.

Well, to get on with my story, time flew fast for all of them with Kyrmen the eldest doing his best to fulfil difficult responsibilities. His mother was wont to depend upon him more and more, such was her nature and such her expectations. She'd so allow the conditions of her life to defeat her that she'd deprived herself of a fulfilling and spontaneous life with her children. Bondage to her past made her flighty and suspicious; one minute warm, the next cold; and she allowed that to take control of their lives!

The one attribute of her love, came through in an errant kind of way, for she cooked and cleaned and saw to it that they were properly fed and looked to their physical necessities as meticulously as if that held the key to their happiness.

She had tremendous energy for scrubbing and washing and cleaning. Despite herself she was driven to do this. Physical work released her from the tensions of a world that she'd rather not acknowledge. Sometimes she found that she could open her heart wide enough to allow the laughter in.

The children would romp about and take liberties with her for awhile, for awhile only. Then with no warning at all she'd shut it again. So what were her children to expect at any given time: the best or the worst of their mother?

When those unguarded moments of love; brief, short periods when her heart enveloped them in its warmth came upon them, they willed it to go on forever. It was as if some lingering memory of their father brought out her ability to love once again. Somehow they managed to get along; and being children, forgave her for her many faults and continued to love her even when they didn't understand her. I sighed with frustration.

Something about her could never accept the circumstances of her life. She'd become a mismatch for everything and everyone around her. No I didn't hate her anymore. I'd learnt to accept her and see why she couldn't allow change to take hold of her.

Some women are like that, some men too, you know.

In the course of the children growing up, she feared Kyrmen's inevitable leaving, feared losing him to another woman. For, as is your custom, you know it yourself, he'd one day, move out to live with his wife; who hadn't yet materialised of course; but who was like an imaginary thorn on her side. Sometimes when he annoyed her, I'd hear her tell herself in an undertone that he'd leave her anyway and thus spoil the mood for herself.

Was she mad? Instead of loving her son simply and completely, like other mothers did she set up imaginary obstacles that held her back from him. But as I keep saying this about her, I wonder if she wasn't inadvertently driven to all this, because she'd been so crippled by her father's unreasonableness!

Do I forgive her? I can hear you ask me. I never knew myself. But I did know that her young ones were as conflicted as she was.

So the knots became more knotted up and family relationships more prone to ups and downs. Days of laughter and love were few indeed as the seasons went by and as Kyrmen grew into full manhood. Kyrmen and his sister, however, never knew anything better so they assumed that this was the stuff of life itself. They never expected anything else or anything better.

Meanwhile, he managed to work the fields and coax out plentiful harvests from the depleted soil. He laboured from morn till dusk toiling for the family, until the land did seem like it had a man looking after it again. After all did he even know anything better? And wasn't he better off out in the fields where he enjoyed certain peace of mind and freedom? And he was young and still had much to learn, so what else to do than to keep himself thus occupied? He'd never learnt to make friends and he was used to being alone.

Most times I'd see his old grand-uncle sitting outside in their front yard just biding his time till Kyrmen got home. Then there would be an evening of peace and quiet, a sense of each other. Kyrmen always made an effort then, to be with his grand-uncle as much as possible. No empty chatter, no wasted words, just two intimate friends, oblivious of the world around. The grand-uncle had taught him skills that no one ever had. Kyrmen bore such a close resemblance to him that even I was often deceived.

I thought he was like him. Yet there was a difference.

I did speak about it before you know. Kyrmen was also clearly his mother's son for, as a young boy he did sense her difficulties and saw her as she was, because at some level they did understand each other after all. He was after all her first born! He knew it only too well after all. He preferred not to think about it.

Phuit, phuit phuit! I didn't want history repeating itself, but I couldn't control the unaccountable forces of destiny could I?

He'd become stubborn like her, as mulish as she was, and though I hadn't yet seen to what extent it could rear its mulish head, it was there all right. It broke through at certain odd moments when he'd refuse to talk to his grand-uncle. Oh it was painful to watch the old man. You must understand that to many of you, these matters that I speak of were never so obvious. I was the one who'd known them and seen them for a very long time; too close to them not to perceive this submerged, complex state of affairs that showed itself through the woman's erratic behaviour, Kyrmen's stubbornness and his sister's flightiness.

The grand-uncle, of course, had lived a different life, separate from them. When he came to stay, he'd had his full share of life's fortunes; misfortunes as well I suppose. He'd married young; but sowed wild oats as a truck driver, ferrying cargo of all kinds from one place to another. His wife had remained faithful to him all the same. But he'd never been able to help his escapades. They were exciting, they were fun, he found women challenging too! On his retirement he finally stayed put with her in rented lodgings. There'd been no children. When she died as is your custom, he came to stay with the niece whom he'd rarely visited.

Kyrmen became his lifeline to a world that he never knew he would want so much to be a part of and would have to leave whether he wanted to or not. Never having had children, never having bothered much with family life except to show the world that he had a wife somewhere, he realised only too late what he'd missed. And he felt close to him, as close as if he were the son he never had.

Being a river I was so close to them too as to be able to distinguish even their individual breaths. To you all looking on from the outside nothing was visible. You never knew what went on.

Only I knew it all, and the old man now; who'd become uncannily sensitive to the world he'd taken for granted far too long. But all this Kyrmen never knew. Only I knew the hidden stories, the secret depths, the mix-ups that unhinged relationships, the concealed lies.

But I was never able to intervene on anyone's behalf at all; simply because I was a river to whom no one paid any heed.

So what about the women in the family then? I'd hear the clatter of tongs and *tyndong sliew-ding* and listen to the sound of Kyrmen's mother blowing hard on it to make the fire spark, so the rice would cook; instructing, instructing her daughter animatedly in the ways of the kitchen. As if all happiness rested on that; on the cooking of food, on the shine of pots, on the kitchen being cleaned till the planks were bleached white, on repeated attempts to domesticate her daughter.

The daughter was the assurance that life gave her, since I see you people set great store in the birth of a girl. As if a girl is the answer to all your problems. *Phuit*! She felt, as I know how many of you women feel, that she was the key to the future, for her offspring would be the grandchildren who would carry on the clan name! The sister was made to understand that she was the strength and support of the family. Kyrmen would leave, her mother philosophically and misguidedly told her, when he'd found a woman for himself. What were they to do then?

Only *she* could see to it that the rest of them would be cared for. As if that were true indeed! Why was she saying all this?

And what about his sister, what about her? She showed utmost unwillingness to understand those words young enough as she was. She could be as stubborn as her brother if she wanted to and equally given to eloquent silences as her mother!

I'd see the smoke rise in the air, sniff the penetrating smoky punch of *dohtyrkhong* hanging on the *tyngier*, as the beef that she bought in the weekly market crinkled steadily over the smoking fire. Life seemed settled and normal then if one were to judge things superficially.

Till the time that Kyrmen fell madly in love, that is; just like that; as easily and as hopelessly as that. He was taken up by a girl whom he'd seen in the market several times before. She was rich; she was beautiful and had more than she could ever want. But Kyrmen noticed none of these.

Her skin was smooth, fair as porcelain. She knotted up her hair in a becoming, maidenly kind of way. Her heels were carefully scrubbed and painted with lard every night so that they were tender and soft as the *kait syiem* banana. She was vain about how she looked. The tassels of her *jaiñsem* played attractively around her ankles and her gait was alluring and inviting. None of these were important to him. It was her soft manner that deceived him from the first.

She was unlike any girl he'd ever met, so unlike any of the women in his family. He imagined talking to her and communicating many things to her and starting a wonderful life with her. The poor untaught boy! He was as simple as she was vain and as stubborn in his love. And it suddenly became clear to me that I was as powerless now as I was sometimes powerful when circumstances favoured me, and made things move my way.

Something about Kyrmen also caught her eye early in their acquaintance. It was the same way that he tugged at my heartstrings a long time ago when I first saw him. She didn't know one thing about him. She merely fancied him; for he was truly different from other boys; even she could tell! She assumed he was the one for her. So she went out to conquer him, the wily female. Could he have withstood her wiles, introverted, untaught fellow who'd never spoken to a girl let alone understand anything about a woman's heart!

I felt as helpless, as vulnerable and as weak as an infant. I knew it would never last. But could I do anything? Could I have offered Kyrmen anything better?

He'd run small errands for her, spend useless hours in her company and do anything she asked of him, absolutely given over to her. This chit of a girl, whose name isn't even worth mentioning, worked her way steadily into his heart, such that Kyrmen seemed crazed with love for a girl whom he hardly knew. He was her slave alright; abjectly so, disgustingly so.

One day she found that he was only the son of a peasant, with no land-holdings to boast of. She was more shocked than disappointed for it was galling for her to find out that she'd been deceived, as she believed. Her outrage knew no limits for her pride was shattered.

I for one was gleeful for I knew that she would drop him just like that, as quickly as she'd pick him up; and then he'd be free again. I hadn't expected him to be so heart-broken. There was meanness in her that came with youth and admiration easily garnered. She recovered quickly from her surprise all right and actually boasted to her friends that her trophy was all theirs if they wanted, for she had no use for him now. She discarded him like an unwanted *jaiñsem* and picked up another whom she thought was more suitable for her.

3

To cut a long story short, since there's more for you to hear, Kyrmen was, as they say downriver – left high and dry. His mother did not make it any easier for him. She'd known all about it from the *lorni* nosey parkers who seemed to make a living out of the misfortunes of others and who made her feel betrayed.

I felt chaotic vibrations like heat waves, burning my insides. I was shot through with so much pain and concern that I found myself doubling up in agony. I was furious and upset. If the family's reaction could do this to me what would it do to Kyrmen? I knew that there was no strong reason for his mother to oppose the girl, but as I told you her heart was like a clenched fist, tight with all kinds of unresolved emotions even now. Not only had she not allowed her love for Kyrmen to express itself simply all these years, but she was now wont to listen to those good for nothing *lornis* who fed her with superstitious fears on sons' choice of women.

She'd been uneasy the whole week and it seemed that the clenched heart was ready to burst. What was it? What was it? I asked, bursting with curiosity myself.

In a dim kind of way she was taking responsibility for the many things that had happened. She actually expressed impatience and grief over her own limitations! But those moments that gave me hope went away just like that and I found that she'd driven everyone away by an even heavier silence.

Kyrmen had shut himself off completely from his family; switched off from them in the same way that his mother would push them away when the moods claimed her.

It was then that he began frequenting one of my more secluded coves again and again, to fish; maybe to get away or just to be by himself, to nurse his hurt in solitude. He was like one wounded, surviving on animal instinct. He didn't have to be like this at all I tell you.

He stumbled upon it on one of his dazed wanderings when he felt that he couldn't as yet go home too early in the evening and face his family's accusing silence all over again. I saw it all I tell you, the family's inability to commiserate, their inability to communicate; their emotional barriers; their, what your young people call, hang-ups.

The mother did want to ease the silences, but she was untutored in the finer ways of the heart and was too used to giving up easily, too used to letting go. So, I thought, what had to be had to be. Kyrmen and his family locked in a conflict of emotions; each one like the other; each to their own misery. The limbo was one of their own making after all.

Anyway I saw to it that he was never disturbed. These coves tucked away from common sight were a prized part of me that I kept hidden from the sight of ordinary people. They guarded deep stretches of dark waters hidden in shade, the impenetrable, unpredictable *thwei*. They seemed dangerous to you because I made it a point not to make them accessible to you. *That* I made sure of. I loved these stretches and allowed my soul to mull over their hidden depths, so that my flow there was definitely slower and my waters denser. This cove safeguarded such a *thwei* and as I told you I saw to it that he would be safe there.

It was a quiet refuge from the confusion around him. Remember what I said? He was as stubborn as the mother from whose womb he came and as he made the cove his stronghold as well as his fortification, he withdrew completely into it as if it gave him immeasurable solace. He'd withdrawn so fully into himself that he seemed to have changed overnight. His face became stamped with lines just like his mother.

I wept for him for I knew he wasn't as far gone as his mother and could somehow be retrieved. Who would do it, however? If she couldn't then who? That set me thinking about her too. Who'd been there to help her?

These were also the *thwei* in which the *puri* lived for they were secluded from humans, away from the noise and clutter of your villages. They were river folks, the only folks that I was close to; companions of my darkest nights, instinctive by nature and endowed with magical powers that gave them great perception. They'd seen Kyrmen but left him alone, sensing his anger and hurt.

But how was I to know that he'd attract the attention of a young nymph who saw right through to his heart-ache and conflict, and sympathised deeply and wanted to take it away from him. I knew then that he could never escape her. My heart sank; not only because she was a *puri* but because of the circumstances that impelled him to her. These were those water nymphs, gentle creatures that preferred to spend their time following the lives of the humans who frequented the riverside. They're as tenacious as death, once they have them in their hold. They never intend harm ever, but they love human folks, oh they love them and always want to be part and parcel of their lives at whatever cost.

There was no peace for him now. His grand-uncle tried to shield him not so much from the others but mostly from himself; tried to counsel him but got no further than a few sentences. But he'd seen it all, known it all. He too had been a young man once before so he knew and thought that the moment would pass. He'd underestimated Kyrmen. He thought Kyrmen was like him!

Kyrmen's sister had also grown up a lot by now and was beginning to see what life was all about. She was puzzled and wanted to ask him, probe him for answers but he turned away from her too. How could he be so foolish?

He guarded his hurt and confusion stubbornly. He began to neglect his duties, neglect even his grand-uncle. I'd see the old one wait up for him almost every evening. But could he do anything about it with one foot in the grave already? Though old man that he was I'm sure he knew. And what would a river like me know about such things? His mother had to know. She was a mother after all.

The pain and enforced silence had a strange effect on Kyrmen. Days stumbled into weary nights that tagged on to the other days and the other nights unendingly on. Chores were perfunctorily done and most times now he headed for the seclusion of this cove. This Kyrmen was unlike other boys. He'd never flirted, never had fun. He'd taken life too seriously. And that one failed relationship felt like the end for him. So he nursed a pain that for other boys would only be an unimportant milestone in the transition to manhood!

The *puri* was a good one. Not one of those possessive ones who brought destruction and pain and violently broke up families. Not one of those your story-tellers make too much of. No, she was one of the other kind, loving and gentle and prone to watching humans with curiosity and warmth. She watched over him patiently, hopefully, gently coaxing him to wake up from this uneasy stupor.

A few weeks, a few more days and Kyrmen began to respond to her in an instinctive kind of way. He was never conscious of her presence but the cove gradually became a sanctuary that seemed to communicate with him in a wordless, pacifying kind of way. She'd been able to cut through all his pain, communicate to him as no one ever had. But as for now he was still unconscious of it.

So he began to stay longer and later, until he fell asleep one evening soothed as never before in his young life. He had a dreamless sleep that refreshed him but the moment he woke up in the morning he was quick to wonder what his family would say.

What could he do? Could he tell them, could he expect his sister to share in any way in any of his hurt or his pain or even this strange experience of sleep well completed after many, many months? He'd never opened his heart to them ever in his young life. Nor had they either; and it seemed far-fetched to do it now.

I saw it all, this struggle within, the unease inside and the, what do you call it, conflict that splintered him and left him vulnerable to other outside influences as it were.

Finally, he did what he was duty bound to do. He went home as he habitually did, but before his family could cross-question him he stubbornly made his way back, to work in the fields. He didn't bother to wait for his usual morning repast. He was only too eager to be away from them. Already I could see that he was fully estranged from everyone now.

Was this a new Kyrmen? I asked myself, or was it everything that had happened to him in the cove, or was it that he sensed that he was on the threshold of something new, something never experienced before, something quite within his grasp.

He worked hard that morning. He couldn't understand why his feet ran like a deer and his hands flying like the wind.

When evening came he thought he'd return home but found his legs taking him steadily back to the cove. He slept there that night and then found himself there again the next night and then again, and yet again on the following nights. He didn't seem to need any food nowadays. He'd cut himself off from his family and was so distant from them, so far away even from his grand-uncle that he no longer talked to anyone now. But his heart was light; light as the breeze that seemed to be blowing through his entire being now!

And now wonderfully peaceful days turned to soothing nights in a complete reversal of what it was like before. The nymph had prepared him well, so when she began to appear in his dreams she just fell into place as naturally as if he were expecting her. Perhaps he did? I can never tell. But I was saddened by the turn of events. Not that she was evil. No not that. I was saddened because I'd seen too many of these relationships end up tragically and painfully. There was always a sense of waste in them But as I told myself who was I to predict such things? Who was I but a passive bystander anyway?

Could I have made any difference?

So from dream to reality is but a short step I should say. The nymph had shed all timidity and had in turn won his trust. She soothed him and calmed him with her presence and did everything to relieve him of the pain that she must have felt bruising him inside. Not until he'd accepted her unquestioningly and completely, however, did he ever learn that she wasn't human. But he took her as simply and as naturally as that. His dream world had already been filled with nightly visions of her so when she came to him in person, he was fulfilled.

She'd tamed him in a way that his mother had never been able to do.

No questions were asked at home because what would they say to each other anyway? I was impatient but held my peace. As for his mother, she'd never had anyone teach her many things. And when she did have a husband who could have helped her, he died on her. If at all he was the one who'd betrayed her! But then again she never understood what she'd missed, so there it was . . . the past catching up again?

And though my story continues in water and in land I want to linger on with the old grand-uncle for awhile. He formed no part of what happened in Kyrmen's life yet I knew that he was there all the time. Too old to help him, he hoped for many things for the boy. He gave him understanding and in his own way gave him his blessings for a future that he could know nothing about.

And so I saw that the grand-uncle, the grand-uncle knew without seeing, understood without knowing and accepted without understanding for he too had once been young and he'd loved many women. Who knows, he might even have wished for Kyrmen's freedom, a freedom that could never be had on land? Who knows, perhaps he identified with him and even saw himself in him? Only he knew but he was unable to reach out at all.

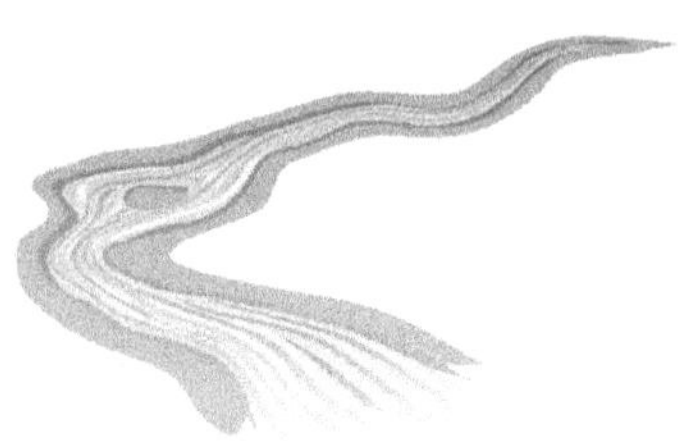

CHAPTER FIVE

Kyrmen Skhem continued to live a parallel life for a long time. Daylight found him ploughing his mother's fields and night caught him descending to a vast watery universe that I hesitate to describe to you in detail.

I may repeat, much much later what he inadvertently described to his mother; but we'll see about that, we'll see; for I don't trust you one bit. Once you hear about it you'll move the heavens to destroy it and not rest till you do. I don't want you meddling into Kyrmen's affairs with your selfish, probing, mocking curiosity.

This other universe that I'm privy to is by far a better one, far removed from all of you and that's why, I think, Kyrmen fitted in so naturally. It was a world where life's substance was not only water but something else besides, something more satisfying to him, something that he'd never encountered in his life on land with other humans or even with his mother. And it was here that I found that his stubbornness and increasing mulishness vanish bit by bit. When he was up there he was a different man. Down here he was another man altogether, so given to loving his family so devoted to his wife and children. He was completely spontaneous; no walls to isolate him.

And as easily as that his old self sloughed off.

At first, his descent took place only at night. But slowly, a whole new world of possibility, of love and family and friends began to hold him back more and more. He was loathe to leave his wife after that, for this she'd become to him, especially now that she was expecting their first child. And she was everything to this boy, for boy he will always be to me, the one who always needed the understanding and love of a mother. If his life on land was full of darkness and unresolved conflicts, this other life was steeped in the mystery and fullness of a love never known to him before.

And so mornings I'd wake up feeling him deep inside me, like a fish darting forth to its destiny; mobile and quick, dashing here and there, knowing that he'd tarry for yet awhile, for what would he go up for anyway? His grand-uncle, his only friend was now reduced to a weak old man not far from gone, unwilling to recognise anyone, locked in a past that'd conquered him forever.

And in a world of intricate, underwater caverns moistened with layers upon layers of marvellous mysteries, his life unfolded as wonderfully as never before. It seemed to him that the *puri* folks whisked themselves here and there in the split of a second. And he'd actually begun to move like them. In this new home time belonged to a different dimension altogether and he was fully alive to every moment of it. He savoured the peace and smoothness of this new life where he was appreciated by wife and family and where the things of the other world had fallen away. He lived like one of them and took care to keep the memories of his other life away. But somehow they persisted.

And in this vast, vast expanse of water Kyrmen never ever realised that he'd been away for a long, long time from his mother and family above because they'd been pushed aside in the daily ritual of a routine so full with his *puri* wife.

2

And then it happened. I felt it; like the jump-start of a foetus in its mother's belly sending shock-waves to her whole body. I too felt the sudden twist, the hitch in my belly, as Kyrmen's quick remorse, his sudden recollection after a long time, of a family above, spewed out whirling, confusing emotions that unsettled him and skewered my balance. They twisted him and tormented him, and left him with a tremendous conflicting yearning to see his mother, his sister and his grand-uncle once again. He seemed almost to have become his former stubborn self again; caught up in this indeterminate desire that rapped him and reminded him of things best left behind.

This boy was truly his mother's son; incapable of letting things be. He must be forever at it, digging up things that are best forgotten. Didn't he realise that he'd lived a full life here? Had the boy never grown up even after the birth of his children?

His two children were as alike as two tadpoles; one was a girl and the other a boy. You'd never have been able to distinguish them from humans; the only difference being in the turn of their feet, aligned backwards, heels placed in front and toes turned back. He'd tell them stories of his other life but never before was he ever seized by this feeling of nostalgia for what used to be hearth and home. He became restless and his face no longer glistened with happiness.

What his *puri* wife feared most had come to pass. She sensed that he wanted to visit his family up in dry land and there was nothing that could prevent him, not her love, not the children, not his watery home, nothing whatsoever. It gnawed on him day and night but he could never get himself to tell his wife.

Over time, he'd changed alright; he'd become more open, more able to give himself to others and that too freely and spontaneously, less given to those unhealthy silences that marked his days in the other world. His *puri* wife had hoped to make him forget his past so that he wouldn't ever want to go up again. But somehow, somehow he couldn't help being his mother's son and his yearning for her was indescribable. Could anyone keep him back?

She knew immediately of course. But it was only when their young son expressed a desire to see this land of his father's so different and unreal and unimaginable, that he finally confessed his heart's desire to her. You see, he'd been telling his children stories about this other world and he'd stirred up imaginations and wishful thinkings that he was going to realise for them, no matter what!

Kyrmen had promised his children that he'd take them up to their grandmother. He cooked up excuses for wanting to take them to their human grandmother and repeatedly assured his wife that whatever her fears were, they were unfounded. He pestered her. He coaxed her. He told many more land-stories to his children until they too chorused their father's desire in other ways than one!

Kyrmen was a loving father and gave his children the best of what he had. And these were memories so close to his heart that he simply couldn't deny them the one opportunity they would ever have of going up to visit their relatives on land. And taking them up now, was his heart's desire; if his love were to be truly fulfilled.

What was he about? Distance had made all memories of his land experiences a little hazy. And besides, he himself had changed so much. He'd lost that old propensity to those sudden moods of unreasonable anger and of wanting to be alone all the time. He'd gained the confidence of an independent man well able to handle any matter whatsoever. He'd found happiness, so to say, and had a family that, he felt, would be the envy of everyone up there.

But didn't he ever realise that they might not be acceptable to all? Or was he trying his best to forgive and forget? I never knew. I never would.

What was to be had to be. Kyrmen finally managed to convince his wife that he'd go, but only for a short while and that the children would be safe with him. He promised her again and again and again that he'd not stay too long and left with her instructions ringing in his ears.

In deference to her wishes, Kyrmen of the deep waters, paid full attention to his wife's words before leaving. She in turn, I knew very well, was as apprehensive and as concerned as any *puri* wife would be. But to go into this would require too detailed a digression for which I have neither time nor patience.I want to get on with my story about my Kyrmen.

So when he reached land he'd have to prepare for her coming for she wanted to join him too. She never intended to let him leave her like this. She never trusted his family, least of all now; but she was a *puri* after all, who never trusted land folks. Kyrmen Skhem knew this only too well. As for himself, he was not too sure how his mother and sister and grand-uncle would receive him but, he felt he knew in his heart that he'd not be unwelcome. He didn't of course share any of these feelings with his wife.

Neither could he understand himself at all: why did he want to see the old world so suddenly? Why indeed, and that too after such a long time? Why did he feel this insistent call of the land?

As he speared upwards, he recalled his wife's words again. After he'd been with his land folks for exactly a week, on the appointed morning, he'd have to begin by sweeping the path leading to the house very, very scrupulously indeed. And then, if he really wanted her to come, the house would have to be spic and span, not a pot out of place, not a bed unmade, not so much as a grain of rice left unswept or a speck of dirt to be seen.

And after all this was done, he must see to it that the *synsar*, the broom be hidden in the house somewhere completely out of her sight. If she were so much as to get a glimpse of it, she'd never be able to meet him in dry land ever. The *synsar* was the enemy that could sweep her back to where she came from. Many of her kin, she told him, were forced to retreat to their watery home at the mere sight of a *synsar*. But she refused to tell him anything further.

About the children, they'd only be able to follow her after a day or two. But they'd only come disguised as creatures of the world outside. He'd recognise them immediately; *that* he need never fear. It'd be up to him, however, to convince his mother that they were truly his children; otherwise she'd never be able to see them in their real forms. And if she didn't believe him then she'd be the one to lose for she'd have ruined all opportunity of seeing them as they truly were.

So simple indeed, thought Kyrmen. What was all the fuss about? Indeed he himself seemed to have changed completely now that he was about to leave. Even I couldn't recognise Kyrmen. He seemed impelled by new desires that were actually old ones that nestled deep within: he just couldn't put away his folks on land, so easily; despite the magic of a *puri* wife.

He'd never been like this before, in a hurry to get there, in a hurry to show his family to his folks up there, in a hurry to go and then come back. He seemed to have mixed feelings, of anticipation and excitement, of trepidation and hopefulness and yet, there was something in him that I couldn't quite understand. It troubled me so much that I felt disturbed, truly disturbed deep down within. I cringed at my own powerlessness but I couldn't help myself, foolish river that I am! Foolish me!

He said his farewells to his family and took his wife's carefully packed *ja sngi* wrapped in the succulent leaf of an underwater lily. He crossed over quickly, his heart beating with anticipation. If he could've grown wings he'd have flown up in no time at all. But he steered himself so skilfully to the other world that he reached dry land in no time at all.

He rested on the riverside rocks that were so familiar to him as a young man, to eat his midday meal. He felt so hungry all of a sudden. Memories flooded back to him and I saw him almost choke with emotion. Nothing seemed to have changed. The path was as it was before, twisting its way to the village, quiet and without a sound. It looked as if it were expecting him.

But I lonely *wah*, lonely river, observing things all by myself, saw that he'd suddenly aged. There were lines on his face never visible before. His hair seemed discoloured and his eyes had lost their shine. Was it the sudden dryness of the land air that he'd been away from, for so long? Can he handle the present? I wondered, finding myself involuntarily wanting to protect him as always. Whatever had happened had happened and I couldn't stay the strong beat of my love for him, as powerful as the currents, as unchanging and as unbeaten as the course that I'd steered all these years.

But then a chain of events took place that shook my entire being. And as always I watched helplessly, uselessly, hopelessly, tearfully by.

The moment he opened the leaf, everything turned to sand! Not a grain of that succulent rice sowed in fields that ripened in the temperate regions below, was left unchanged; even the leaf shrivelled at his touch. No matter, he thought, unperturbed. So unlike his earlier land self! He was back. He could always eat later. He only saw the path to his mother's house and ran shouting impatiently for her as if nothing had happened, as if no time had passed at all, as if his mother too had changed, as if, as if. . .

3

At first no one recognised him. Then his mother, who always expected him back, must have somehow sensed something familiar about him: the mere fact of his presence perhaps, or a mannerism here and a gesture there, for she immediately hugged him instinctively. She knew in the only way that a mother would. It was as spontaneous as that and it warmed my heart as it did Kyrmen's too.

Oh she'd missed him alright. She too had aged recognisably. There were lines on her face that had never been there before. She looked beaten and old but not unlike her earlier self. I hadn't bothered with her all this time so I'd lost touch completely. She seemed to have matured somewhat. At least her heart had learnt to make place for her human grand-children now; I sensed a certain loosening, a certain tolerance that was never there before. I hoped that this reunion with her son would bring her round to accept what couldn't be changed. He was waiting for just that after all. But, after all, only time would reveal all.

So I had to wait on the sides with my river nymphs, after all.

His sister already had a brood of her own. She was no longer the simple girl that he'd left on land a long while back. She looked stressed and I must say her whole bearing reflected certain physical strain, for she seemed preoccupied with her growing family; it showed on her.

Her face was covered with *ñiang mong*, the speckled brownish mask that all you women seem to be afflicted with after you've borne a few children. I didn't see her husband and assumed he was out in the fields or in the market. She'd filled out into a woman. Nothing good-looking there; just someone ordinary who'd mothered several children but who handled life's twists and turns sensibly. She was the exact opposite of what her mother used to be. I sensed conflict, yes, as in Kyrmen Skhem of the old, bygone days, but also a quickening sensitivity that told me a different story about this woman. Surely she'd outgrown her own mother!

I hadn't met up with mother and daughter all these years. Who knew what went on? Who knew if there were disillusionments and vanished dreams even here?

However, at that moment, the genuine happiness that the reunion brought to mother and son and sister, even though their grand-uncle had left them a long time ago was well worth seeing. I cried my heart out for the son who yearned to renew his land connections.

It didn't take too long for his mother to understand that her son was already a father and that too, of strange creatures. She recoiled from the thought of them, but pointedly and in an uncharacteristically single-minded way told herself that they were her son's children; also reminding herself that they were her grandchildren. She tried to question him; tried to find out more than what he told her, tried all means to prise out information about his life away from her; but Kyrmen kept mum.

He was guarding a valuable treasure, his present life; happiness that he never ever wanted land folks to snatch away ever again. He'd only tell her what everyone else thought they knew.

After meeting those on land Kyrmen realised that nothing had really changed. If at all it was he who'd changed. The mere fact that he hadn't been with his mother for such a long time freed him from the responsibilities of a relationship that his sister must now be battling with, he now assumed. Were her loyalties torn between mother and husband? Not that I knew anything about her husband; but if I were to go by appearances he must have been a good provider for the children looked happy and well-fed.

Perhaps the onus had fallen upon her to mother her mother now?

It was only a matter of time before Kyrmen put forward the request to his mother to allow him to bring his wife to her. She'd expected something like this, but was surprised that the request should come so soon. She saw an earnestness that now touched her to the quick and since being re-united with him again, she'd even told herself that she'd put up with anything from him even that which seemed to be her son's foreign ways; so she agreed.

She too was affected by the change in him. He was animated and communicative and wasn't afraid of showing his emotions. Perhaps, hoped the old woman pathetically now he'd come back to her! She'd missed him, true; and had subjected herself to a half-hearted kind of introspection since he left years ago. She'd take her son back, no matter what; she'd yearned for him and missed him so much. I wanted to seize this unfortunate woman and tell her not to let go of what she now had! Then I saw that she was actually making an effort to make place for her son! Had the separation taught her something after all?

So plans were made for the *puri* daughter-in-law's visit. But, in as much as his regard for his mother seemed to have grown a bit more during this visit, so was his love for his wife growing even deeper now. And so Kyrmen worked hard for his wife's visit. He looked forward to her coming as much or even more, than he did to see his own mother.

Anyhow, on the morning that the *puri* wife was to come up, there was an air of expectancy about the house. She dutifully helped her son to follow instructions carefully; she actually helped him clean the house thoroughly. Not a speck of dust on her furniture, not a pot out of place whilst the kitchen gleamed spotless. The windows and the two doors, in front and at the back, were thrown open in welcome invitation to the visitor from the other world.

There were niggling doubts within me, despite it all. But I cast them aside, I caste them aside. I spat out *phuit, phuit , phuit,* to ward off anything bad.

So was Kyrmen assured of his mother's earnestness and so did he go to the water's edge to receive his wife with a light heart. He'd lived simply and naturally down there. The exhilaration of coming up on land and meeting his mother and sister and looking forward to having his family up here had thrown him off guard so to say. And so he was as vulnerable as vulnerable can be, because for the first time he'd opened his heart to his land folks in true *puri* fashion.

But then there it was and to be fair to all, so was his mother equally vulnerable. She was making a gigantic effort to receive someone she didn't quite understand or know too well; and knowing her, it was admirable but scary.

She could be rocked either way.

His mother sat waiting for them. I could see the effort this cost her, but as things went she did try her best, you know! And at that moment I wanted to explain to her that those pangs of jealousy were only part of being human and that apprehension about meeting an unknown daughter-in-law who wasn't human at all was also normal.

I could understand her. But could she understand her *puri* daughter-in-law? I couldn't help asking myself. It was troubling but also touching to watch her fight her inner fears. The separation had done her good, subdued her somewhat and made her less insensitive. Maybe she now understood what her son was to her; maybe, maybe. . .

She was full of trepidation and her thoughts were clouded as she sat waiting for them in the wooden veranda enclosing the house. The moment she sighted the two of them, walking up the path to her house she stood up to greet the *puri* wife. She was struck by the unworldly beauty of the other woman and hard hit by the realisation that she'd lost control a long time back. What was more, she immediately noticed her son radiating a certain glow of happiness. She'd never seen him like this before. He was unrecognisable to her.

Simpleton that she was, she was puzzled and didn't fathom what it was about the two that simply excluded her. A familiar shadow swept her face as she reacted to the sight and I felt the coldness in her heart. It was the coldness of anxiety and dread not of rejection, however. Thank the gods it was only that!

Meanwhile Kyrmen's wife never felt at ease at all. She was beginning to regret the visit. But I couldn't see her too well here in dry land. I could sense that she was clearly waiting to be alone with Kyrmen so that she could make her doubts known to him. She already felt constricted I could perceive that. But Kyrmen was oblivious of the tug-of-war that went on within the two women. He helped his mother lay out the meal. It was a subdued one. No proper conversation took place. Kyrmen never noticed. He was too happy. He was too filled with joyful expectations. At the end of the meal, his wife dutifully helped her mother-in-law put away the leftovers in the *duli* and clear away the table but was unable to help her clean up the *sohkhiewja*, rice grains that had fallen flat and sticky on the floor. That required a broom.

His mother's nerves were stretched taut. She was also irked by the other woman's tardiness, displeased that she didn't offer to sweep the floor or she must have been shocked by her inverted, backward turning feet unnoticed until now.

Who was to know after all? Even I felt bewildered by the turn of events. It distressed me to see both women unable to meet each other even halfway.

What was wrong with that woman? I asked myself. Was she going to throw Kyrmen's good-will away? But I calmed down immediately for I realised that it wasn't that she didn't try.

It was just that they were completely different from one another; their elements would never mix. Kyrmen wasn't any help either; for he wanted them to be friends the moment they met each other. How could that be possible? It was far too soon and too rushed. His mother felt that it was only the other day that he'd left her. The knowledge that her husband had had a stressful relationship with his mother stared his wife in the face. So how could they simply put the past behind them so easily?

There were awkward silences in the kitchen. To be quite fair his mother did try you know, but land and water are two different elements and the two women were simply stuck in their elements.

Of course, there was a patchy conversation now and then and his mother talked about this and that, but found that she could never steer the conversation round to inquire about the *puri's* family or her grand-children either. Something had caught her tongue. And the *puri* too found herself equally tongue-tied; so alien was she to land life and manners.

Then Kyrmen's mother who was always so meticulous about cleanliness, realised again that the kitchen hadn't been swept clean. Kyrmen had never felt it necessary to tell her that the broom was the enemy of his relatives and friends from the underwater world. She excused herself as she got up to fetch the broom. She was itching anyway to move her hands and legs for she was woman who was used to working, after all. After awhile, she came back to the room; broom in hand, relieved that she wouldn't have to sit inert like a fool idly facing her daughter-in-law the whole afternoon.

In the blink of an eye Kyrmen's wife was nowhere to be seen. Kyrmen was stunned. He'd have followed on his wife's heels had he not been expecting his children; he must have also remembered what he'd promised them. Nevertheless, he ran after her but was too late to stop her; nor did he try to follow her down to their water home because he must have been confident that matters would be settled with his wife later, when he got back.

When he came back from the water's edge, he didn't seem too disturbed. Maybe he'd expected this to happen, maybe he was too caught up with the anticipation of showing his world to his children and maybe he hoped that his children would change everything; maybe, maybe, maybe . . . there were too many maybes. I never expected him to be so composed, however.

His mother, meanwhile, was taken aback by the turn of events. She attributed the sudden flight of her *puri* daughter-in-law to her dislike of land folks. And just like that, old imagined hurts and prejudices began to surface slowly, one by one. When Kyrmen tried to tell her, explain things to her she was wont to disbelieve him, but found him so changed, so convincing, so bent upon making her understand him that she wanted to listen after all. She wanted to heed him no matter what. Perhaps things might change after all! But I knew that things were not what they seemed to be.

About Kyrmen, only this I knew very well: he wanted so much to show his children his world, the one that he came from. Just a glimpse of it would be enough, he thought. And he expected his mother to accept them unquestioningly even as he continued to assuage her fears and apprehensions about their mother. That was why he was patient with her. He was waiting for his children to come.

What further proof to her of the full life that he now led down there?

4

A day or two passed by and I could see Kyrmen biding his time for his children to come up. But it seemed an endless wait. I watched Kyrmen waiting, waiting impatiently for them. More than anything what I realised later was his earnest desire that his children see this other world. He wanted them to share in the familiarities of his early life. They were his. His to show the land-world and his to take wherever he wanted to.

Was he too being possessive about them now, wanting too much from them? The same, like mother like son pattern being repeated again?

Anyway, let me get on with my story before I too begin to choke with too much, too much emotion. This was the season of toilsome chores before the winter cold, when days become shorter. I was familiar with all its different sights and sounds. The rice had to be husked, the chicken coop repaired, the bamboo fencing changed, the pigsty cleaned out completely and odds and ends to be seen to before the frost fell. The harvest season was already upon them and the storehouse had to be swept clean. Leftover grain had to be stored elsewhere to make place for the new crop and the clouds seemed to be gathering in the heavens as if threatening to bring down those soggy autumn showers that spoilt the yield. The yard echoed with frenzied activity every single day.

At mid-morning, Kyrmen relieved his mother of the husking when he saw how hard she struggled to pound the rice with the heavy wooden *synrei*. A truce had been consciously maintained. His mother tried her hardest to put behind those things that she didn't understand but her mind was still awhirl with the unresolved events that brought in even more unresolved thoughts. This time I couldn't blame her. She tried to behave as normally as possible but I could see her cracking up, for the rupture had been forced again no matter what.

Could it be bridged?

She'd been told about the children's coming but never about the form that they would take when they came. So she expected children like anyone else. Blame it all on Kyrmen this time I told myself. But the blame was neither hers nor his.

It was just that the two worlds could never come together. The elements would not intersect, they refused to meet. That was Kyrmen's tragedy after all.

He rested for awhile, after he'd husked a basket of sweet smelling rice. He intended to continue with the work a little later. His mind was elsewhere, wandering the cool recesses of the waters below, distracted by thoughts of his children's arrival; he'd lain the *synrei* aside. He never expected his mother to take it up from him to continue with the back-breaking work.

It was as if she wanted to use it for something other than just pounding the grain, for she plunged it almost savagely into the *thlong*. Perhaps she was wrestling with her thoughts and pounding them away trying to clear her head, still pre-occupied with what had just happened? Had it been in earlier days she'd have withdrawn moodily and crazily into herself. But her human grandchildren had succeeded in wedging open the door to her heart bit by bit and it was virtually impossible to shut it again. Even then I heard her doubts and her fears ticking away within her.

And then it was that she saw from the corner of her eye, two flying dots making their way towards her. At a distance she didn't at first know what they were. But as they drew closer, she made them out to be two brilliantly-coloured butterflies flying in her direction.

They were handsome creatures. Their wings reflected the colours of the sun and they were flying together in perfect unison, perfect freedom, like tadpoles skimming the shallows. There were many that day that saw them. They marvelled at the way they suddenly flew out of the river straight there, (later no one could even tell the exact location where they came from), as if they knew where they were headed for. They'd created a sensation with their magnificent colours. And when they reached there, they chose in all their innocence to alight on the *synrei* that their grandmother had put aside as she raked up the grain in the *thlong* before she would start on the pounding again.

They rested there with wings folded up. Yes I'm sure you know. They'd seen their father and they'd made straight for him or perhaps they recognised immediately the grandmother of their father's stories. In their innocence they thought they'd be accepted as they were. To them things were as normal as normal can be and they expected to change into their normal selves soon after.

But things happened so fast that even Kyrmen didn't realise anything, anything at all until it was too late.

They were probably planning to rest for awhile till they got their breath back or till they were used to the dry air. But the old woman, whose heart was still taking time to recover, who was not quite herself as yet, thought in a moment of characteristic pre-occupation only with things concerning herself, that they were useless insects who'd lay eggs on her grain.

She could never imagine that they were the grandchildren whose arrival everyone was waiting for. Beauty had always escaped her and their exquisiteness at this moment was of no consequence.

I was appalled.

She raised the *synrei* to scatter them. Then, finding that they didn't fly away, slammed it expertly and unthinkingly down upon the two exhausted butterflies. They never anticipated such unexpected violence, least of all from their own grandmother.

They were too tired to flee.

Kyrmen only saw the exhausted flutter, the hopeless twitch. He cried out in shock and disbelief, and rushed to them. Stricken by the sudden turn of events all he could do was rush up to them.

When he picked them up they were as still as death.

What about his mother then? What about her? She was at a sudden loss for words, at a loss for anything to say. It took time for her to understand that they were Kyrmen children. She'd never faced such a situation before and somewhere deep within her I knew that she knew that she'd committed an unspeakable act. But she was also in shock.

Even now I don't know what she must have felt because Kyrmen was who I saw, who I wept for. Did she understand the seriousness of her loss, of Kyrmen's loss? I've never bothered to find out.

Kyrmen cradled them in the palm of his hand carefully, lovingly. By this time a crowd of curious friends and acquaintances had gathered around, for the cry was a terrible one, not easy to forget, this terrible, terrible cry of animal pain that haunts me even to this day.

Blind to anyone else out there, Kyrmen was like a hunted being. And when he turned to his mother I saw how she'd destroyed it all completely, in a moment of importunate preoccupation, selfish preoccupation only with things that mattered to her.

On his side, there were no recriminations, no tears, no show of emotion; nothing at all; just that: sheer emptiness. He felt tired. In the dreadful clarity of his loss he saw what he shouldn't have done; but he was past any kind of regret; past any show of emotion.

In wanting to see their father's land that which they'd heard so much of, his childhood home, they'd lost everything. No! He'd lost everything. He had only himself to blame. It was a terrible moment for him; of intense grief and sudden realisations; of insights and bitterness, of helplessness and despair.

He was back where he always was.

Then, even as he was making a visible effort to control himself, he stepped up to her and told her in a stiff, even voice in the presence of all of us there, that as long as the bamboos were alive and standing, in *his* cove, hidden by the sheer granite wall and the twin pines, she should know that he would still be alive. There were no explanations. That was all. That was all I tell you! I wanted to hold him close, erase those bitter memories and bring his children back to life.

But I was stopped in my desires when I saw that nothing could be done anymore. His mother had done it all; in that one moment of inordinate preoccupation.

He turned back to the water cradling his loss in the palm of his hand, as empty and as dazed as when he first met his *puri* wife.

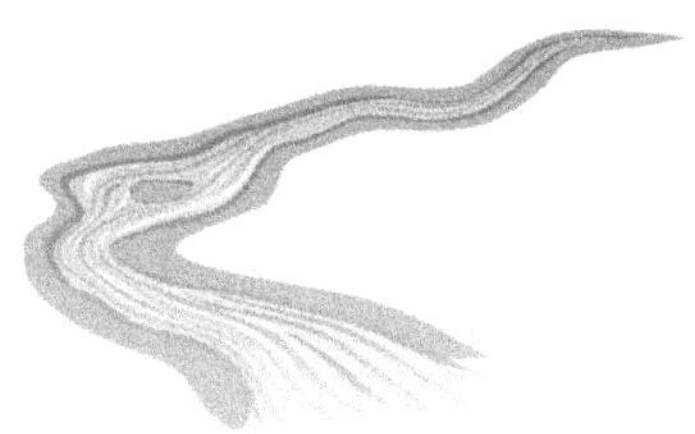

CHAPTER SIX

Meanwhile Bem had disappeared for a long, long time and now that I'd caught up with her again, I heard that she'd eloped with her butcher that night after everyone had gone to sleep. She lived with him and bore his children. But unlike many women in a similar situation she remained faithful to him even though he turned out to be, a sadistic, alcohol guzzling monster.

The morning tea she served him from the time she moved in with him was actually *kyiad* poured out in a cup. Her life seesawed between misery and short-lived happiness, wife-hood and mother-hood. She bore him a child after every weaning. It was unending frustration. And then suddenly one day, at the age of twenty three she was left with the care of five young children.

What do you think I felt? Could I have turned the days back? Could I have done something to make her see reason, bring her back to her senses, pull her in and hammer her with my waves, my way of speaking to her, but you've seen haven't you, how insensitive and indifferent these young ones can be? I was as agitated as only a mother can be. What's more I saw it all; I saw it all and simply had to look on.

Her husband suddenly had had a fit in one of the *pata kyiad* and his friends had, soberly and dutifully brought him back more dead than alive. Perhaps he'd been over performing, zigzagging his way through the hearts of many customers and cronies and women again and again? He barely lasted through the night. She'd been with him eight long years and reaped nothing but babies and misery and a battered sense of being. She'd also changed from being young and hopeful to being weary and wary and cautious.

There was no one she could turn to at this moment. And how could she expect her mother to help? She'd wooed a new husband and just succeeded in getting him to marry her before he changed his mind. He was several years younger than her and she had to keep an eye on him for he was the only man she'd met who had a regular income.

He was a "registered contractor", whatever that was! Her gold earrings flashed with passion and pride, as she talked of buying land to her growing children. "Finally", Bem's mother would congratulate herself, "after so many false starts!"

But I was unnecessarily filled with anger. False starts indeed! I could only churn round and round, churning up a whirlpool of emotions that would have sucked in anything, anything in my agitation. I felt hot and cold all over. I knew the signs well, murderous anger for the mother, even more so because I couldn't touch her.

But when I looked back later my anger evaporated because what else was there for her, what other option did she have but to grab this timely opportunity? Her daughter had been given a lead and she was on her own, or rather had been on her own for so long. So why shouldn't she learn to sort things out on her own?

Her mother was doing just what was best for herself and for the other younger children who were her responsibility. She wanted happiness so badly and Bem was already well into womanhood; well able to take care of herself. It was bewildering to me alright, but I couldn't get myself to be really angry with her anymore.

So Bem had no one to turn to. Her other siblings were in various stages of growing up and some of them as small as her own little ones. She couldn't possibly ask for anything from her mother. But she'd been born to fight in order to live and she found that she had to fight even harder for her children; a survivor who pitched in everything she had even though she lost out in the end.

She was unlike the other girl, Lili the foolish one, the one who never learnt the one who sold her soul to the *sor* for no good reason at all.

These were girls who needed help. I was the only one intimate with every detail of their lives and I churned with the knowledge, knowing, knowing so well that they'd never hear me, nor heed me.

Because I saw it all and felt the frustartion that crushed them, I hovered close to them even though I was separated from them by this watery body from which I could neither reach out nor touch them. I'd slam myself again and again on my banks as mad and as crazed as when my currents were being churned up by the torrents of monsoon.

"Of what use is all this, this show of emotion?" I asked myself heedlessly, for I knew that our lives were destined to cross; and because I couldn't keep them off, couldn't put them out of my mind, and yet couldn't wash them down like I did those others who fought me off, couldn't even dispose of them as I've done many, many unfeeling others; for these were different don't you see? They needed a mother, not a punisher.

2

As for Lili, as soon as her father died, she felt a sense of release. She left the family that she was staying with and took up lodgings with an older woman, experienced, who knew the *sor* inside out, who'd lived and worked in many *sor* households, who was her adviser in so many ways, who promised her a better life; a fatter income; a new lifestyle; "permanent stay in your very own lodgings from where you'll go to work in people's houses on an hourly basis, a *nongtrei kynta*! "No one will dare infringe upon your rights then", she informed her.

"What rights", I couldn't help replying even if only to myself "do these girls know about?"

"That'll bring you more money than you'll ever see in your whole life stuck to one household", she assured the foolish girl who was now a woman. "But you have to give me a small percentage. Better still I'll stay with you and you look after my food and shelter".

She was shrewdly domineering and unlike the rest of them, a *sor* offspring, a natural who affected superiority over all the others. She made it her business to pry into the lives of all these girls who came to the *sor*. But they always kept her at arm's length for they sensed that she was craftily meddlesome. Lili was the only fool who fell for her wiles.

She felt fortunate that someone so like *a* mother had come to stay. Perhaps too the woman did want to be a mother to her but hadn't known how? She used Lili's earnings as if it were her inheritance. I watched helplessly from the sides, watched her father's efforts go to waste and saw the way that she was headed. I heard her father's words of caution and pictured his concern. And I re-lived the old man's fears. But could I do anything to stop her I ask you? I found that she'd slipped through my fingers as easily as the wavelets cascading away from me.

And before that one year of "freedom" was over, of all people, Lili attracted the attention of one of the locality boys who lived off what the streets could offer him. She was plain, remember, and as dull as a donkey. All he had was his youth and to the silly girl he seemed as handsome as a god. She passed him everyday on her way to work. He of course, had certain necessities and he knew, street fashion, that she had money stashed away, for he'd often seen her go to the bank.

There were gossips on the streets all too willing to discuss her with him, carelessly and without sensitivity, something they did about everyone, just to pass time; how much she had, how much she was worth. They even predicted to each other how long one could live with her before the money finished.

He of course listened carefully. He must have planned it all out for one day I saw him make his moves. He had indeed planned it out meticulously.

Her earlier employers had instilled in her a sense of caution in money matters. "The bank" as she often told her lesser-informed *nongkyndong* relatives "where girls like us working in the *sor* keep our money is an impregnable fortress".

She talked about it in a special tone, imitating the mistress, and they were always awed by her *sor* wisdom. I knew her well though and saw right through the very words that she used. She was only parroting her well-meaning employers.

So I saw him begin wooing her without as much as a second thought. When I first saw him at it, I was ready to tear him to pieces like that Khrawbok. My body itched with repulsion and loathing. Had the time been right I'd have trashed him like a piece of rotting flesh, food for the *mukur* that grew fat on trash like him. Had he ever come near me when my waters were murderously overflowing he would never have lasted a second.

I would hear him mumbling to himself when he was drunk or whistling cheerfully when he was in a careless mood, and saying something about "visiting her and surprising her with my presence so that she doesn't fly away before I've made my scoop."

The brainless, silly girl was flattered beyond compare when she found him waiting patiently for her even when she returned late from wherever household she was working. She'd make a fuss of him and cook extravagant meals for him, giggling flippantly and unashamedly hanging onto his every word and gesture. As if he was the only one for her.

That familiar feeling of rage again, hot and cold, hot and cold; rage, for the sake of the father who believed he'd given his daughter a life! I wanted to avenge him but had to wait for time to take its course. It was frustrating, as frustrating as when the rains didn't come that year. And I felt as impatient.

He was a superb actor. He surprised even me After an evening of successful moves, I'd sometimes get a glimpse of him in the dark walking briskly back in a leather jacket and boots that he'd recently acquired, his cap scrunched down on his head. I got a sense of the fellow's smugness and confidence as he strutted like a cock, single-mindedly. It was nothing to him that he was playing with a woman's life. It was everything for him to feel the thrill of manipulating a woman, before he snatched away her money, the object of his pretended love for her.

I could never look her in the face for a long time after, for I'd see her father's face reflected in her forehead and I'd feel distraught with shame. Such men were these girls attracted to.

Didn't they ever learn?

Hardly any decent time passed before he unceremoniously moved in with her. After that, it was history repeating itself as in the case of many *nongkyndong* girls like her. After a short while he kicked out the surrogate mother. They were too evenly matched for one another. One had to go. And it was the surrogate mother who finally left. She was beaten hollow by the other *sor* natural.

Happiness was short-lived in Lili's house after that; and the evening laughter began to get wilder and wilder over days of celebration with those other drunken street friends of his.

I'd listen uneasily unable to sleep for whole nights altogether afraid for the foolish girl who'd given herself to such a man. And whilst the drunken brawls increased I was stricken with some kind of unknowing yet knowing fear. If I could have nightmares like you do I would have had several such nightmares.

All I felt was this cold clamminess that froze me completely and slowed my movements. I was sluggish when I should have been active, tied to an imprisoning element that taunted my uselessness.

Despite everything, despite whispers of him being a casanova who flew twelve flags brazenly all at the same time, a *rang khadar lama*, Lili loved him.

She wanted to love him. And as her stomach grew bigger, her gold earrings disappeared first, then the thin gold chain and then the gold ring that she'd made for herself when she was working with the family that employed her at the beginning.

Dimwitted, unteacheable and foolish as a *kada* or was it *gadha,* as those Hindi-speaking migrants swore, I thought. I didn't want to see more of her. I was fed up. But I couldn't shut my eyes from what I saw.

There were too many illusions in the lives of these deluded girls; confusions and mix-ups that could easily be avoided, but who was to tell them? They insisted on following them. They didn't care if anything happened. They obstinately wanted things to happen their way.

She'd become thin and undernourished and as a result her work began to suffer. Her longstanding employers noticed the change and the inevitable slackness at work. She was more absent than present nowadays and they needed to have their homes cleaned and dishes washed and other work done. They were losing their patience as I mine. She'd be absent for many days altogether because she was too weak to go to work and already her employers were beginning to grumble about her. And to top it all, even when she had to stay home sick in body and spirit, she couldn't turn to anyone at all.

She wasn't blind to the man's failings anymore; but she was unable to get out now. She herself had allowed him to push her into this kind of submission. But she preferred not to think of it as that. Maybe she feared that if he left her she would lose credibility in the eyes of those who craved for just this kind of security, this so called "married" state that she herself was in. So she tried her best to salvage whatever self respect she had left.

She seemed to play the game with surprising dexterity, however! When he was violent she was passive, when rude she was patient and that gave her a certain undeniable strength. I never expected this peculiar mix of the extraordinary and the pitiable together. She wasn't a push-over yet her pliancy and defencelessness were her strongest defence, a passive kind of tenacity that was tough to overcome.

And I noticed something about the relationship that kept me wondering at the strangeness of it for a long time afterwards. The rogue was actually finding it difficult to break her hold on him.

I unwillingly had to admit that to myself. Was she in an absurd kind of way in control? That was a question only he could have answered for he found that he was unable to look her in the eye anymore.

One day when she was at work, he must have been driven to it; he stripped the house bare and left her finally, to fend for herself. She found herself alone, in an advanced state of pregnancy. And for once when she came home that evening, she cried her heart out. But not for him, strangely enough; there was a sense of her own part in the whole affair. She cried with a sense of relief for what could no longer be.

However, alone, Lili's defenses easily crumbled. I saw her that evening, walking hesitantly up and down the path leading to her husband's newly-rented lodgings, undecided whether to go to him to plead with him to come back to her or not. She never plucked up enough courage though.

Her baby was due any day now and nothing was ready; there were no blankets, no *jaiñsop* and no baby clothes; not to mention money. She was now distraught. She'd been nurturing an irrational dream, naively expecting to hold on to the man through the unborn one. She'd hoped for too much, for too long and couldn't let go of him just like that.

And as for me all I was concerned with was the unbirthed one, swimming in an unbirthed universe full of water, struggling for life in its mother's womb.

This was the girl who imagined herself to be strong and clever, able to take on the *sor* and all its complexities! The girl who'd imagined many fine things about herself; who'd put off her *nongkyndong* sensibilities for this! What was this? I could never understand them when they reached this point in their lives. I was filled with that familiar sense of failure and anguish, of cold unfailing anger at these men and shame that these girls had no self-respect whatsoever. And these situations were as irreversible as the flow of my waters, southwards to the plains, onwards to an uncertain future.

Why does it have to happen so? Why were their hearts so hard, so unmoving so difficult to change?

One morning, Bihthei the *nongtrei kynta* who'd struggled to keep herself above board, doggedly staying away from the temptations of the *sor*, met her. This was the girl, I'll concede, who'd succeeded in striking a balance between her *nongkyndong* background and her *sor* employment through sheer grit and tenacity. Now she was determinedly working towards re-fashioning her life for the man who'd approached her. He was a trainee recruit from the police academy nearby, who'd soon get the coveted stripe pinned onto his shoulder.

She was lucky she'd met him almost by fluke for she wasn't attractive at all except for that dimple on her cheek and she wasn't going to let go of him by any means. Unlike Lili, adversity had honed her inner resources. She was shrewd almost to the point of being cunning. This was disguised by an immense capacity for laughter and fun. She'd managed her life single-handedly up to the present moment and there was a sense of achievement now that she was close to attaining her objective of marriage.

Bihthei found Lili almost crazed with pain; labour cramps that came at short breathless intervals. She was beating her stomach and flailing her hands wildly at the same time. She paced the cement path outside her house up down, up down, and at close intervals I saw her stopping to press her back against the wall as if pushing the pain out. There was a glazed look in her eyes.

This was her first pregnancy and she was absolutely ignorant and unprepared. There were no women friends, no female relatives, no informed or ill-informed midwife to help ease the pain. She was up on her feet when she should have been prepared for the birthing. She'd cut herself off from her sisters and now she found herself alone, helplessly pacing, pacing the path with an exhausted kind of energy.

Bihthei immediately saw the anguish written on her face. One more look at her and she saw her vastly protruding stomach. What could be done? She was already aware of this foolish girl's plight. She'd warned her several times before, but to no avail. Lili had cut her off completely and refused to even speak to her.

But Bihthei, who was also capable of strong loyalties, concerned for her obstinate, foolish neighbour, characteristically took matters into her hands. She rushed her to the hospital where Lili almost immediately delivered a baby that was so underweight, it was as small as a *tyndong tympew,* the bamboo cylinder where betel leaves are preserved, and as shriveled as a baby monkey. I waited breathless with anticipation, instinctively reaching out for a child that I hadn't even seen. It survived for a short while: only to give Lili the satisfaction of knowing that she'd birthed a child, her very own. It was a girl. My mouth tasted sour.

After that, Bihthei somehow managed to get in touch with the other sisters in the village. She had to use all her powers of persuasion to convince them to come quickly and come they did even though they were not familiar with the *sor*. Then they took her back despite her strong protests.

I overheard Bihthei relating all this several years later, to the neighbour who'd moved into Lili's house after she left so suddenly. And then she revealed something else that chilled me even on that hot afternoon.

3

Lili had had a child from the new husband she acquired two years after her return to her own village; but he too had left her for another, smarter woman who also had plenty more to offer him. She'd lost her youth, her only asset, and she'd become as wizened as a middle-aged woman. She'd begun to look like her father, but without the wisdom of experience that gave him dignity.

Her concerned sisters had given her a small plot of land on which, with what they called "government subsidy", she was able to construct a pukka dwelling for herself. She expressed her desire to rent it out, so that she could return to the *sor* to work there. But her shrewd sister, the younger one, threatened to evict her even from her own house and disown her completely should she ever leave again. Lili stayed on unwillingly.

She slaved in the fields and ploughed the earth in the hot sun. Her only child, her sole companion would play in the dark, red earth while she toiled under the burning sun. But who was to know what plans she had for her child?

For, one day without telling anyone, Lili, whose hair had become brittle and discoloured by the scorching sun, who'd shrunk to half the size that she once was, whose youth had been stolen away so cruelly, came back to the *sor*.

"She was unrecognizable" said Bihthei, "until I heard her voice". I strained to hear more of the story, lapping up all the details of Lili's present life; and then I heard, "she's even brought her child to work for a family she hardly even knows!"

Their voices grew fainter and fainter and I strained to hear more.

"What unfulfilled fantasies - - -" I caught the words.

I almost washed down a part of my left bank trying to follow them. Then it was that earth pushed me back gently, nudging me back on course, reminding me my place.

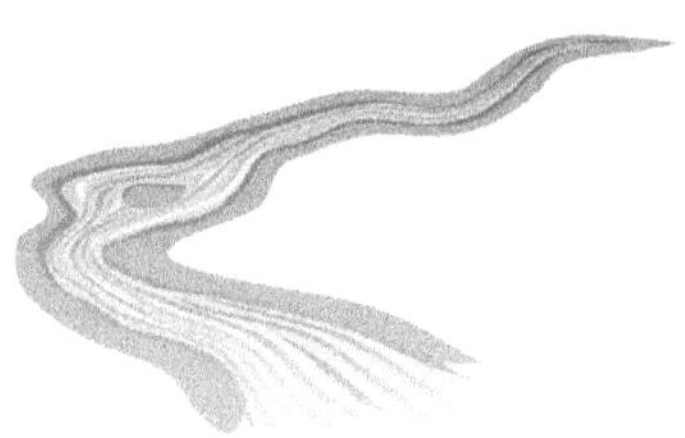

Chapter Seven

And now to catch up with Bem. She was one of my favourites from all the others. But what do I tell you about her? She wasn't perfect, she wasn't anyone whom you could admire. She'd lost her youth when I met up with her again, exhausted with unending troubles, wound down like a tired old woman. The one thing that tied her close to me was her unflinching loyalty – to anyone she loved – that was what gave her standing in my eyes.

She was special, in a certain kind of way. Was that why I felt an instant bonding with her even when she was a child?

She'd never gone back to her mistress after she moved in with her butcher. One would think that she'd forgotten her. But it wasn't that. She was just ashamed; ashamed of her actions, ashamed of the mess that she'd made of her life. This was the mistress who'd listened to her childhood stories; the one person who'd given her a listening ear; the woman who understood because she too had had her share of rejections.

And she found it difficult to forget her.

In the early days of her relationship with her butcher she was too happy to have someone who loved her so much. But when her eyes opened to the man she was living with, whose next child she was always expecting, naturally there was remorse and regret that she'd never listened enough to her mistress, to have learnt how to deal with life.

The other family members had never been close to her. But all the same, she'd often think of them nostalgically. She had no home to go back to. And that was the only home she had ever had. Those were in the early days of a strange, bigoted relationship with her butcher; sad days for sure. As I watched her, I wanted to hold her, speak to her, tell her that her future was as much hers as mine.

When things got bad with her butcher husband she preferred to keep to herself and would invariably think of the mistress who'd been like a mother to her. This was a game that she played with herself and they would always bring in those dreams at night: they were a mystery to her.

They were like visitations from another world altogether. They calmed her and freed her and enabled her to go about her work as always. What work, however, did she have but that of a woman whose life consists of the dreariness of keeping a large family fed, washing her family's rags and hanging them out to dry, scrubbing pots and kitchen floor, facing a drunken, rotting husband day in and day out.

Her butcher would often come home drunk as a fish, throw his food at her, curse her and try to hit her, and then deliberately, provokingly call out the names of his past wives one by one.

"They're definitely head and shoulders and everything else above you", he'd spit out cruelly at her.

She didn't mind all this, didn't mind being hit; he was always too drunk to be able to touch her anyway. And I saw even then that what kept her going were those extraordinary dreams. I envied her for them; but they were also what made her self- sufficient and yet defenceless. And I also felt a deep insecurity within her that communicated to me and told me things about her that weren't easy to put away.

But I was tied, as I told you, tied to my useless, water body. I was ineffective when I was most needed. She of all the others I loved most because of what I saw inside of her. But I couldn't help her in any way and I hated myself, hated myself bitterly for being so utterly useless.

These were not nightly occurrences, however. That's why they were bearable. Sometimes he'd come home sober, pockets overflowing with the day's earnings and spend memorable evenings with her. It wasn't too bad at first. She rather enjoyed the sense of being married. And he was earning good money. She did get to buy many things for herself too you know, spoil herself for awhile. Not that he was totally involved in ill treating her all the time. Those were moments filled with something like happiness however short-lived.

He loved her then as he'd loved all his other wives and would spend time with her when he took off from his friends. I silently celebrated them with her even as I saw what her life now amounted to.

2

Things went on smoothly for a year or two. She was made to put her old name away. "*Phuit eh* your name reeks of unpleasant memories", he told her. 'I'm going to call you Dasuk. That's the name your mistress gave you, isn't it ?"

So Dasuk she became but stayed so only while he was alive. After his death she was forced by everyone, friend and foe alike, to drag out her old name and put it on again.

No matter. I saw the old Bem resurface, the Bem who could keep herself happy under any circumstances, the Bem who never expected much from life. But I'd also see her trudge to the market reeling under realities that weighed her down, that were so sudden, that were a living nightmare, with a houseful of children to feed and no one, not a soul to turn to.

Her old name did bring back a hunted past. She was powerless she knew it now, but she forced herself to go on. Somewhere from within came a strength that fought off all those disappointments.

Does a name matter so much? I've always wondered about this. You attach so much importance to a tag and the marvel is that you believe in it so implicitly. Bem couldn't shed her old name because everyone around her forced her to use it all over again. They forbade her to forget it.

Did it matter to them? Not one bit. But they continued to insist upon it and corner her all the same. It didn't make any sense at all to me, not at all.

And so a new, old life began again for her. "So you're the infamous Bem", I'd hear the scornful remarks.

"Bem turned Dasuk turned Bem again! How's that? How does it feel to come back to yourself?" jeered a neighbour who'd interestedly kept herself informed of the relationship and followed it through to its very end.

They added other names to the stock of names that she'd acquired. She was the butt of everyone's cruelty. They never knew her as I did; never saw into her, never bothered to know who she really was. I cared for her like a mother, a sister, as much as any of you humans, didn't you see? But how could she ever know that "Bem" was only "Bem", a name and nothing else? That she was above it all?

Who would tell her? Only the dreams could tell her, but did they? Did they?

Bem never answered. Her sulkings took the form of sullenness now. Merely went on with her work. She'd had to fend for herself, for her five children, now that her butcher was dead. She'd found work with a woman who'd been her husband's business partner.

This woman sold *dohsnier* and *dohjem* entrails at a make-shift stall located in a curve in my banks. She was strikingly good-looking and tall where Bem was plain ugly and dumpy. She had a husband who partnered the business with her and this made her a supercilious witch. I despised her, the miserable wretch, despised the way she scorned Bem, despised her for everything that Bem never had.

She took great pains to ridicule Bem, call her names as if she were her owner, as if this would give her a better sense of life, as if this would drive away whatever it was that was irking her about Bem. Bem, I could see, stayed on only because she had to. Like the dumb, diseased animals she'd cared for in her mistress's house, Bem felt as useless as them and as maimed. When once too often the husband approached her for favours, Bem threw the *dabor*-full of *dohsnier* entrails that she was hanging up, at him and walked out with only her self-respect intact. She hadn't been paid that week and wasn't going to be. From somewhere deep within her she'd summoned that old self again and she walked off happy that she'd done what she'd done.

She surprised herself, surprised even me, that she could be tough, but tough she was as days became lonelier and nights even more so. But I was never convinced by her tough exterior. I'd seen her when she was alone trying to keep herself above all the misery, so to say. I saw her having to fend for herself ; it took everything she had.

I was overcome by a terrible sadness. I'd never been as moved as this. I'd never seen the depths of her wretchedness. I felt her desolation deep within stirring chords never stirred before. But no one knew. Only I knew.

I wept.

Not anyone in sight to help her. Her mother who'd grown prosperous, tried to help her for a short while, but her daughter's youthfulness was a threat and she was afraid. She had a young husband to reign in and a position to keep in society, now that she was a matron with some respectable standing, so she had to pretend ignorance and left her to fend for herself.

And I wondered so much how personal equations change even between mothers and daughters as I bleakly saw both sides of the picture and learnt even more, the ways of your world; learnt to see as I struggled hard to accept that mothers too seek a life for themselves after all.

"After all what's there in having five children with no husband?" justified the mother who was keenly trying to reign in her new life.

"She can always find another one as I did. Let her try hard on her own."

She could never understand that she and Bem were two different persons; two different women with possibly different aims, separate purposes, diverse ways. Could you blame her either, that she would never have the ability to understand, or even to know?

Bem was sick at heart. I too was sick, sick by it all. I was at my wit's end. I thought I could have done something, offered help of a kind, but the only help coming from me was this – only this passive hearing – to intensely private thoughts spoken aloud only to herself, at night when she sought me as what she must have considered me to be: unhearing and unfeeling; unknowing and never sentient.

It was untrue and unfair for I was there beside her all the way. But she was never to know. How could she?

She was down to her last *suka* and her youngest was sick. A week had gone by and there were no openings anywhere. What was she to do? Her children were sometimes given food by concerned neighbours as she too had been fed in the past.

"But for how long" she asked herself hopelessly, "can I depend on charity like this?"

When that hot summer morning the children were unnaturally restless and hungry she boiled a few cups of water and sprinkled crumbs of tea leaves and sugar into it. That was all she had for them then. They had it with the rice that a kindly neighbour had left for them. Then she sent them out to play whilst she nursed her ailing one.

The fatigue and worry must have had its toll on her for she drifted off for awhile, only to dream those dreams again

3

When she awoke she felt strangely refreshed and strong.

"Was it the dream?" she kept asking herself.

Why didn't she tell me something then? Why didn't she speak? I wanted to know.

There was something playing on her mind I knew at once. I also knew that she couldn't quite pin it down to anything because she called out to her children and uncharacteristically, told them to stay inside, not to let anyone come in till she returned. There was a tone in her that reminded me of the old Bem.

So, the old Bem hadn't completely disappeared then!

She was behaving strangely. And I was alarmed but more than ever I was overwhelmingly concerned.

She went off with no real objective I could see, only the blind urge to go somewhere. Like the Bem of old.

I trailed her with my river-sensibilities determined not to let her out of sight from now on. She walked off in the direction of the marketplace as if impelled by some inner impulse. But I could see that she only had a vague notion of somewhere to go.

Then she turned in at a *teem* counter, pulled towards it as if by some extraordinary force that had propelled her there. She cut a ticket for herself, with a number written plainly there. Was this the number that she'd dreamed not very long ago? I almost burst with curiosity and watched with growing concern. But I hit a blank wall there.

"Is this it?" I thought aloud.

No one paying any attention to her? Did it was always have to be as I always saw it? Bem alone against everyone else?

And when she returned empty-handed, her last *suka* gambled away there was nothing left inside or outside of her, none at all. She'd scraped the bottom and found nothing there. There was nothing. And I kept trailing her, appalled and aghast at what she'd done.

But could I have even offered her any money? I didn't even know where these things came from.

I too was empty of all feeling. If only she knew I was there, I'd give her everything I had.

But I was separated from her as east from west, as sand from granite. No opportunity ever arose for a proper talk and I was frustrated by my limitless limitations; my fumbling flow that could never be taken for language of any kind; my dumb association with the human world; my seeming woodenness.

Even though I sometimes felt that she responded to me in a strange kind of way they were too few and far between for me to set too much store by them; but as I told myself again and again I could have been very wrong. Maybe she did try to tell me, talk to me, hoped for a response from me! I threw myself hard against the boulders in my course and felt the anger rise. I felt my waves swell up with an indescribable emotion that imploded within and caused me to swerve treacherously to one side. It split me up and smashed me but it was also a moment of intense revelation.

I could never leave her.

But I also saw that the fight was still in her.

Somehow they got through the day without anything to eat. Her youngest was very still, very sick. The burning fever raged on. Evening found them drinking the same brew as in the morning. Thankfully her eldest two had gone off in the morning as they usually did, to work as daily helpers in the house of a well to do neighbour.

"They've already been fed", she half-consoled herself "and won't be back till they've had their evening meal.

But what about the other two not to speak of my youngest," she worried aloud. "What do I do to help him? I have no medicine, no food, nothing at all."

She was cornered; driven to desperation. So was I.

What else was left now? Only to lock her three younger ones inside once again and go off to the *teem* counter just to see if her dream number had struck cash. She went off just to ease the tension within. Mine was at bursting point.

I trailed her and kept her within seeing distance. It was incredible how I kept at it! But I did it because I cared so much for her, too much in fact to allow her to slip away.

Besides, that was the only hope left. She didn't know what she would find. But she went all the same, sustained by the same hope that had kept her afloat through all the troublesome years with her husband.

Even following her at such a distance I felt the unnatural throbbing of her heart, the unnatural racing of her pulses. When the slate that displayed the winning numbers informed her that she'd won twenty rupees, she stared blankly at it. Thankfully the bookie was too busy with his other clients to notice her stunned reaction. When he thrust the money into her hands she almost fainted with shock.

In a daze she ran home; then stopped. She suddenly knew what she had to get: medicines and provisions for her sick and hungry family. I was also in the race with her running alongside her, for food and medicines; running against time and hunger, sickness and death, pain and frustration; running, running, running exhaustedly with her; running the race for life with her.

And that was how her life began to pick up. A few days later, and as they say it never rains, it only pours, she was offered a job at a teashop serving *jadoh* and tea; also to cook and clean and scrub and do whatever she was capable. It was a dog's life all over again, but it would bring food and money. And she was ready for it.

Besides there were also those periodic *teem jackpots* after the dream visitations. They helped sustain her family. But they visited her only when her children were sick or when she was in need of just that extra money so that she would hit her dream number again.

Strange but true and that was how it was. Bem was too engrossed in the struggle to live to think too much about this strange phenomenon.

Strange for me too, I was totally involved in Bem's life. I wanted to follow the direction that it would take, now that I saw how those dreams came to her and affected her.

Do I hear you laugh at me, ridiculing me for, what you call, my inquisitiveness, that I watched her like a specimen from another world?

Haven't you seen? She had no one, no one at all! You must know that I sensed Bem's mystery connection with that something else that I could never quite understand. And I knew for sure that she stood apart from the rest.

I wanted to help her as I wanted to help all the other girls who came my way.

Her life settled into a daily routine. Her children were growing up and the elder two were, fortunately, on their way to getting an education from a woman who sent them to morning school in return for their labours. Even then providing for her three younger ones was not easy.

Bem slaved from dawn to dusk. There was an energy within that couldn't be contained. She dealt with all kinds of customers but remained detached from all of them. She entertained no one. But even then the men were attracted to her.

"Is it my ugliness?" she asked herself good-humouredly, "or is it that they see me as a inferior victim, someone they can treat with impunity? There's not a man that I can trust and as I've sworn to myself, I'll never take another husband."

So she never entertained thoughts of another man. But sometimes it was difficult to keep them away. They were so insistent, so bent upon getting her, wooing her despite blatant discouragements from her, that she would almost give in to them. She was young. But with growing maturity, she began to see through them and their distasteful play, and could never go too far with any relationship. She never spoke about it, however, for fear of jeopardising her position. So she found herself in many situations unpleasant and rough, again and again.

And she was as helpless as I was. And like me she fought them all off. Something there was that prevented her from taking another husband. What it was she could never know. Nor did she bother to find out.

I knew we were two of a kind.

Meanwhile, the dreams somehow helped her move on. I was happy when she had those dreams. I too felt comforted. The disquiet that I felt for her, seemed to go away when I saw how much the dreams were now a part of her life.

They calmed her somewhat. Sometimes they so completed her sleep, she said so herself, when she would come to sit beside me for awhile, that they helped her get through the day. And they sustained her through all those years.

Sometimes they scared her, scared her with their uncanny insights. But she went on no matter what. The sulks, I no longer saw and the stubbornness seemed to have disappeared. I suppose Bem was growing older and getting to know herself better.

4

And then after a gap of several years, she made up her mind to visit her old mistress after all. This was when the dreams became so persistent that you could say she was dreaming even while she was working, thinking of the dreams even whilst awake.

Such was her state. Such was my state too. I was feverish with anticipation as if this would be a turning point in my life. So I waited for the day. I didn't want to be left behind at all.

She went on a Sunday when the shop was closed, after her weekly clean up at home. It was already evening when she set out alone, nervous and apprehensive. Once there she found her mistress, for that's how she still thought of her, alone. Her children were away with their father visiting their innumerable relations.

Her mistress, *mei-i-thei*, expressed joy and happiness and was visibly moved by her visit. Bem had expected reprimands and accusations but there was none of that. The warmth with which she was received fully assured Bem of *mei-i-thei's* total acceptance. They simply picked up from where they left off.

I overheard Bem later tell everything to her daughter; tell her movingly about the reunion and how she would visit her as often as she could! This was one of those rare times when she was truly happy. They reminded me of those days long ago, when her husband came home with money for her, only for her.

She filled in all the details of her life for *mei-i-thei*, even told *mei-i-thei* about the *teem* strikes! *Mei-i-thei* merely listened and shook her head in wonder. She too had had her dream visitations that were so real she told her, they spoke to her of many things. Sometimes Bem even figured in them. That was when she would have that familiar feeling that they were still in touch; which was in a way strangely true. Much was left unsaid, much unstated; but both knew that these were matters best left unspoken.

I, however, breathed in harshly. I didn't want to understand the need for silence and the necessity to let go. I'd never been able to do it, with my churning emotions, my pain, my anger, my love, my frustrations and I wasn't going to listen! Neither did I agree!

But I found that when I was alone, flowing ceaselessly alone with my thoughts, prone to unnecessary anxiety, I was forced to think of Bem and how she went on despite set-backs and disappointments, ups and downs, seemingly immovable obstacles in her life. That's when if you observe me carefully you'll find me flowing calmly, smoothly by as if nothing would ever sink; controlled by memories of Bem.

After so many years, as Bem was talking to *mei-i-thei,* she began to understand the inexplicable way by which they continued to keep in touch. There was a pattern to them no doubt. It centred around the dreams that she dreamt almost every other sleeping moment towards the end. It was as if deliberate; as if her brain inadvertently declared that it would work overtime and take control when she slept at night. She sometimes saw *mei-i-thei* or some others whom she had known, speaking to her; images and symbols signing to her in a way that only she would be forced to make sense of.

She felt compelled to stay on but finally had to leave. There were no expectations. They understood life's intangibles so well; there was no necessity to waste useless words. If she wanted to come back she could.

Somehow I too calmed down as I began to see everything in that light. I was given a new perspective that I was never even aware of. I felt an unjustified sense of peace.

But, "this is a strange kind of peace", I remember thinking, "for it comes from observing one of you. Never have any of you ever been able to give me this. Will it last?" I then asked myself.

I had only to wait and see, so I continued to flow on, flow fast, flow slow, flow expectantly, flow strongly on.

The dreams continued to be dreamed, especially now that she'd reconnected with *mei-i-thei*. Sometimes when they visited her too often she went to see *mei-i-thei*; knowing that something or the other had happened. Sure enough she'd almost always find something amiss and she'd spend time with her, comforting her so that the years would magically fall off. I made an effort to accompany her always; phantom-like, noiselessly in order to bear silent witness to their conversations.

I was surprised at the change within me; as if I'd mellowed down to match the changing circumstances of Bem's life. They'd talk about the past, the present, acquaintances that they both knew and I saw the mistress's children, now older, wonder at such talk.

They too seemed affected by this person who was at first absolutely unrecognizable, who came unexpectedly after so many years, after they'd almost forgotten her, and who brought with her a sense of the past that brought a certain completing calmness to their mother.

These visits were good for both of them, tonic for all three of us. I was privy to this little world of theirs and didn't want it to stop. But time was running out for *mei-i-thei,* Bem understood this, understood what she would soon lose but could she have done anything at all? After all who was she, just another helpless woman, adrift and alone?

Mei-i-thei's children gradually began to accept Bem's sudden, frequent visits. They now saw the tie that connected their mother to her.

Bem always knew when there was trouble by way of illness or when the family had problems they couldn't handle. That was when she rushed off without a second thought, to help them with their troubles. She was like the daughter who had a special place in the family.

Sometimes the dreams hinted at things concerning herself, her own children. She learnt to pay attention to them even more: they were the dreams that spoke urgently to her in voices and symbols audible and understandable only to her. These were not ordinary dreams. They bore that certain something that could never be disregarded. They called out to her again and again.

In the following weeks, after her first visit to *mei-i-thei,* Bem gained an inner presence, a maturity that came with her ability to accept many things in life. I noticed that business at the tea-shop picked up steadily and she handled customers in a way that she'd never done before.

Her children were also growing up quickly. The older two, a boy and a girl, were doing well at morning school and serving *their* mistress as best they could. The younger two, both girls, were growing up into pretty creatures. They were fond of playing on my banks and I must say that they were truly different from other children. They seemed content with their lot and for now, gave their mother no cause for worry. But educating them was becoming expensive.

There was also her youngest, a boy who was growing up by leaps and bounds, who paddled vigorously in my shallows every morning and who had every intention of claiming his mother's attention the whole day long.

Bem had to look elsewhere to make ends meet. As for the two older children, "they're virtually mine" said *their* mistress.

Bem was reluctant to let go but thought it better to leave it at that. "She's kind to them and makes sure they go to school everyday; no worries there for me", she would try to reason with herself.

It was the other three that troubled her. Because she was illiterate and ignorant and had seen what education could do to people, Bem yearned for an education for her three younger ones. But there didn't seem to be any opportunity anywhere.

I watched her growing anxiety for her children, watched her worry about them and wondered why the dreams never helped her. I felt ignorant, I who always knew what the future had in store for others. But I didn't mind that feeling at all; it didn't make me feel inadequate for I knew that this was a girl who was unlike the others who were so easy to read and to decipher.

Bem lived her life to a different beat and I wanted to see what the future held for her. I followed her, even tried to step into her shoes in order to feel what she felt, understand what she thought, dream what she dreamt.

Sure enough, a year later in an unpredictable way, she was told by a kindly customer who frequented her tea-stall that he'd put in a word to the authorities of a certain school so as to enable her to get more gainful employment as a daily-wage cleaner, on a casual basis. That was something she never dreamed she'd become.

Even for me this was an extraordinary dream come true. A proper income was what she badly needed at the moment and it seemed that Bem's dreams for her children were becoming real enough. They hit the jackpot for her!

5

As things got better they also got worse, a blessing and a curse. She was told to go to the school to be "interviewed" by the principal. He must have been satisfied with her for he offered her the job almost immediately. Also her three younger ones would be given a free education and would be allowed to stay in the school orphanage. She was beside herself with happiness.

As soon as she could, she took her family to the school and found lodgings for herself close by. As she got acquainted with her job, so too did everyone else get acquainted with her name. Even here it followed her. It would never let her be. It stuck on obstinately. She was made to carry it forever it seemed, for it took little time even for the school-children to learn of her peculiar name. She became the butt of their endless jokes and an embarrassment to her own children. She bore it all. For the sake of her children who now had a future. She'd come a long way from the girl with the two thick plats and *nongkyndong* ways.

Just a few months back her children looked to her for their needs. Now in the split of a single second, they looked elsewhere. And they were being blown about by the cruel winds of changefulness; caprice, pride, ignorance, immaturity.

And I felt my hackles rise but my anger was self-defeating. I couldn't undo anything at all impaled as I always was upon my own feelings; feelings that regurgitated back all the time like the currents that swept in and swept out regurgitatingly.

Her name was no longer her worry. It had never been anyway. She could have been nicknamed anything for that matter; it would never touch her now. What hurt her most was the change in her children's manner.

They now treated her almost indifferently. Without being able to get close to them anymore she began to withdraw into herself. The older two were also living too far away now. She couldn't meet them as often as she wanted and she felt that she was no longer part of them, they'd grown so distant.

I felt a despairing kind of loneliness that I'd never felt before. She was groping in the dark. This was the girl who had had the strength to bear her disappointments so well. I slammed myself again and again against the banks.

Even the *dohlun*, the tadpoles, were not as powerless as I was.

She was more Bem than mother to them now. As demanding of her attention as they were before, they were so indifferent now, it verged on the cruel. They'd been transported to a new world that now encouraged them their dreams of a better future and they began to see her in a poor light. She was unneeded. And they grew intolerant of a mother who seemed now to be too full of her own limitations.

Poor Bem, she found she was back where she started: alone, rejected and friendless. Her children had moved on and she didn't belong with them anymore.

She continued her visits to *mei-i-thei* but everyone saw that something was not quite right with her. She hadn't visited her own mother either for there was nothing she had in common with her now. None of her siblings ever visited her or when they did it was always the same story.

They saw her as the stupid one; the one who refused other men because as they scornfully exclaimed, she loved her butcher too much. "There're plenty of men", they told her "from whom you can take your pick".

But she just wasn't her mother's daughter. Men didn't interest her any longer. She wasn't willing to sacrifice her freedom anymore. Somewhere she understood all this in the darkness that continued to prevail.

On one hand the children were doing well. On the other, they'd grown apart from her. Her world seemed to be breaking up into confusing bits and pieces. *Mei-i-thei,* her sole comfort nowadays, was also getting on in years. Bem saw it written plainly on her face.

When I saw her walking back one evening, she'd become like the Bem of those days when her husband had suddenly died on her and left her fending for the children alone. Then, the fight was still in her. Now, she'd suddenly given up completely. There was nothing more to live for. But how could I impress upon her that she was still young and free and still had the world at her feet? How could I tell her that even then she still had so much to live for? That she could still find a husband?

Life had dealt her the cruellest blow of all. It had opened a future for her and her children, giving her an alternative that would benefit them all, but it also took her children away from her. What could she do?

She didn't want to deprive them of their prospects. She wanted to have no part in snatching their future away from them and so she decided to do only what was best for them. She'd simply move away. I was alarmed and worried. Here was she, as strong as ever on the outside but crumbling on the inside.

Who could help her but herself alone? What about the dreams? What about them? She'd speak in despairing tones to herself almost every night now, as she was consumed by feelings of unimaginable hopelessness.

Her children didn't have anything much to do with her anyway, for she was no longer their provider. They'd been "assigned tutors", so I heard it said. There was something impenetrable about that. There was something impenetrable about her too.

Maybe she still dreamed her dreams; maybe she wished to change her life. I wept inconsolable tears. I identified with her in so many ways. Her destiny was mine. It was a cruel, cruel life of abandonment. I felt the same kind of rejection and a similar sense of failure.

So what was there to live for? She began to disregard the dreams even though they were so persistent. She'd already made her options. I knew that; just as I knew every current flowing through me.

She still ran errands for everyone and was indispensible to the school. She was the school itself. But the school authorities were ignorant of their inadvertent backhandedness.

They were her children's savior. They were her destroyer. And no one was aware of this. Why did it have to happen so? I've asked myself this question many times over but never found an answer.

"Maybe there is no answer." I felt I had to accede. But I smashed myself against the rocks again, to fragment my pain, to make it split into a thousand fragments.

She'd deliberately given herself to living a dreamless life, arid and flat even though her dreams were getting larger than life itself. She shunned them and tried to strike them off completely. I knew this because she'd slump down sometimes on quiet, empty evenings on my banks and uncharacteristically hold herself to herself.

Bem was having dreams so frequently they were more than she could handle. They told her urgent things that she knew instinctively but didn't want to heed. They left those messages that she would have listened to at other times. But this time, she'd had enough. She'd blanked off the dreams from her life.

What more was there to know? Her nights became shorter for I'd see her putting off sleep; engage herself in endless unnecessary tasks till she almost always dropped down with weariness. Then she'd give herself over to a night of dreamless fatigue.

One evening she felt strangely compelled to listen to her dreams one last time. They called to her in sleep, crept into her waking moments and even took over the cracks in her life. She went to see *mei-i-thei.* As soon as she knocked she detected that certain something in it and knew. She'd died a year back.

Mei-i-thei's children had tried every means to get to her. They hadn't known about the new job and the new lodgings. Bem spent a vacant evening with them and left without a trace of emotion. She seemed strong to them and they drew comfort from that. They expected her to visit like old times. They weren't to know what went on inside of her. They weren't like their mother, the poor things.

She allowed her life to become a tight routine of endless service to the school. I alone knew about the dreams; what they meant to her, what they told her and how she tried to shield herself from them, to erase them, erase them and turn away from them.

She'd talk to me the poor girl. At such times, I wished I could have taken *mei-i-thei's* place but as I bitterly found out, what must be must be; there are no two ways to it.

What more do I tell you? What more do I add to the story of her life? One night, many many months later, she must have dreamed the biggest dream that couldn't be put away any longer and struck the luckiest number of all; for, when she was absent from work the next morning and the next – this conscientious cleaner on a casual basis – the whole school realised that it missed her and needed her. It sent fellow cleaners to look for her.

Naturally they tried her lodgings first. They found an open door and a crumpled bed. But no Bem. The search parties that were sent out never found out where she went. Was she dead? Was she alive? Had the dreams got her at last?

And as for me, I flow on dreamlessly; unimaginatively grooved to my course.

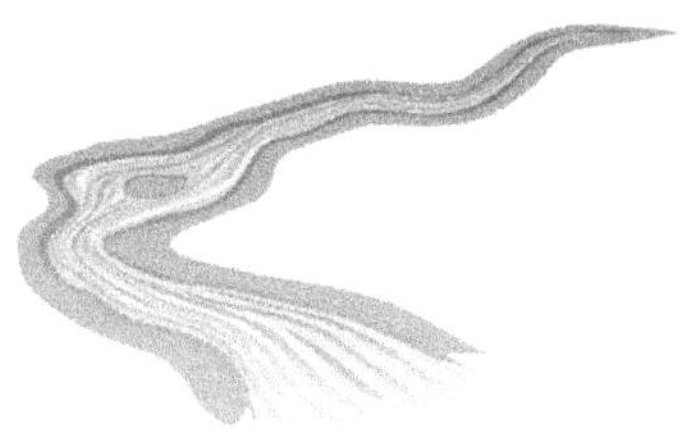

CHAPTER EIGHT

"This *wah,* these *maw, these dieng,* these sticks and stones, these susurrating *'lakait,* banana leaves, will be my witness for as long as I live! Don't you dare point your fingers at me, you off-springs of dogs, you fornicators, you lowborn ingrates, you!"

I strained to hear more.

"You dogs, you *poi-ei,* you daughters of - - - ! Do you think I'm lesser than any of you? Just that my son's in jail you speak ill of him? Does that give you license to ruin my family's reputation? This *wah,* this *wah* bears witness to everything that's happened, I tell you". I noted the anger in her voice, but I also noted a something there that I hadn't sensed before. Was it laced with a deep sense of hurt? Or was it only a show of anger?

I listened again, alarmed and shocked, but somewhat pleased that I was being invoked even under such dire circumstances." But what's she saying?" I asked myself, as I saw her suddenly throw herself heavily on the grass and sob out her anguish to me.

This was Duh Symper, the skinny one, yes, *the* Duh Symper who sent showers of tobacco-laced spit everywhere when she spoke. April Showers, I'd nick-named her a long time ago when, in her vehemence, she sprayed me with April Showers; such was her tongue; such the force of her speech, such the vigour with which she enunciated her words when she wanted to!

No one dared follow her here. She'd stormed out of her tenement like a mad thing, when, as they say, the eagle and the crow whispered things into her ears. She was thin, she was wizened, she was small and wiry, and she'd borne seven sons into this world and had a daughter equally small, equally high-pitched in speech and volume, equally fertile.

The sons were of all sizes, big and bearish, skeletal and sickly, ordinary and talented and who, like their dead father, had plenty of zest for life but very little direction. There were also those who were nondescript, bowed into submission under the full weight of sibling rivalry.

Two had brought their wives with them. They found it easier to stay close to their mother for she was willing to baby-sit for them when they needed her. Needless to say, her constant presence so close to them, gave them warmth and security. So, they'd found houses nearby and moved in next to her.

Another was a bachelor, a rolling stone who gathered no moss. The other, for the life of him, never knew how to manage a wife. His mother had to intervene in all his relationships. For the moment he was alone and, as on several occasions when the relationships failed, he stayed with his mother in a permanently-temporary kind of way. He was jobless. But he didn't mind it, because he would get one when he wanted it. He was an excellent carpenter and much in demand. He also knew that he would get two plates of rice no matter what. So why make that extra effort to leave his mother unless, of course, he got tired of her.

The fourth or was it the fifth son, I can't seem to remember, married "into a rich father-in-law" as it were, who gave him his creature comforts in return for married loyalty to his daughter. He never came back ever, even to visit his own mother.

One was a plumber, thin like his mother, toothless and grimy and tenacious as her. He had five sons but was still aspiring for a daughter. Even now his pregnant wife was getting into the routine for childbirth.

The seventh was employed with a furniture-maker. He was childless and uncommunicative. Only his fingers caressed the wood, speaking-ly, as if searching for wood-nerves and wood-edges, searching, searching for the right places to saw and shave and whittle and polish and shape all day.

So who was in jail? Which one of her sons? I wanted to know; for I knew that this was in some way related to those strange things going on at night.

And as Duh Symper launched her tirade against the world at large, she revealed fragments of a puzzle that I never came close to solving till now. I'd never paid much attention to her before. She was wont to rave and rant, like one crazed when she was disturbed or provoked. Otherwise she was remarkably level-headed when it came to handling her children which was why I admired her. But when I heard her now something told me that she was vastly troubled. This wasn't any petty infighting amongst her sons. I knew it from her voice. This seemed to involve serious matters.

The sons were young and energetic alright. Some had moved off to start new lives for themselves, some managed to stay on, to still hold on to her *jaiñkyrshah* despite their wives and children. Some were hard-working and well-earning young men and some had formed an early habit of eating too much *kwai* and drinking too much *kyiad*.

I guessed that she'd been a victim of what had come to pass. Perhaps that was why she was shouting recklessly at the whole world in general like a cornered animal that was making a last-ditch stand? In a way she was a plucky woman. She was taking on everything single-handedly, the fears, the disappointments, the anger, the set-backs, the apprehensions, the deadly twists and turns . . .

And then it was that she came to me, rushing to me of her own accord as it were, as if to seek solace in me. I knew that she spoke directly to me on that day, that late afternoon rather, when everyone kept away, for fear of her tongue. This was the only weapon she had against the world, the only weapon that could effectively protect her now, it seemed.

She'd brought children into this world, loved them and nurtured them and hoped for a better life for them never realising that they needed something more than just that love. When their father died of a curable stomach disease, because they were too poor to buy medicines, she set her sights only on her children's future.

The moment the coffin was lowered into the ground she vowed that she'd be both mother and father to them. She was small and tough and had had to fend for her family all her life. That was why she intimidated us all. But that afternoon I got the opportunity to know her so well, that I warmed to her as I would only to another river.

I found out after all that she was as defenceless as her children, as defenceless as me but as determined as I was, to live her own life her way. And as I began to understand her better that afternoon, I saw her as she wanted the world to see her and understood her much more. I knew that she had to put up this show to convince everyone that she could take on the world. It sounded daft but it was true. But that's life after all. And we were in it together.

I managed to piece out information bit by bit, about her sons, who were always at the centre of her conversations, who I guessed were in some kind of trouble. But what it was, I didn't know then.

They'd deferred to her in many important matters, but never in the choice of their wives. Even then she'd accepted them one by one, as they brought them from as far away as Mawkynrow in the east and Porsohsat in the west. They were good-hearted young men but had grown up undisciplined. They'd needed their father's firm handling after all, but that couldn't be helped, thought Duh Symper philosophically.

She'd tried all her life to rein them in. But, they'd had to go out into the world to earn their livelihood early in their teens and the world had taught them more about life than their own mother. But out of compassion for her they kept it all away from her. She always wanted to make up to them for all the losses that she thought they'd had to endure with the death of their father. Something had gone amiss somewhere in the bringing up, but she refused to admit it in public.

Let others say what they want. She'd find out herself what was best for her children. They were hers. She passed on favours to this one or that because she loved each one so much. So instead of a disciplined bunch of young men, she had sons, some of whom were worldly-wise, some mature beyond their years, some still in varying stages of growing up. But for one, all of them, however, managed jobs here and there that kept them going. So she hadn't brought them up too badly after all.

The daughter was the precocious one, having had to fill in as mother and younger sister and older sister throughout her growing years, looking after them, looking after them, looking after them, in her mother's absence when the mother had to go out to work in people's homes. She'd matured fast and attracted boys of all kinds. One day, one of her brother's friends fell in love with her and at eighteen found her hands full of the responsibilities of motherhood. In a short span of time they already had four children!

Still I saw that the sons weren't all that bad. They weren't bad at all.

I heard it said by people that Duh Symper had a weakness for her children.

Some of them were failures. Of course they were veritable failures in the strict sense of the word, who'd also never learnt to appreciate their mother fully.

And however much she nagged them, however much she tried to push them on in life, she never had it in her to be entirely cruel to them or to speak ill of them in front of other people the way some mothers do. Would you call this weakness then, in a single mother?

And then I heard her going on about the son who'd "sacrificed his life for *his* people, *his* race, *his* land".

This outburst of pent-up feeling was outrageously deafening. She fell to accusing people randomly, until I thought she'd burst with anger and hurt, so obviously did she make her feelings evident to me.

2

Outwardly she was loud and dramatic; she had to be. But she wasn't convincing to me, not at all. There was an uneasiness there that was wholly unlike the tough woman I knew, who'd learnt to fight the world to feed her children and herself. Was she hiding behind a barrage of invectives?

Even as she raved and ranted against everyone, I gathered that it was the bachelor Nang who did odd jobs for people now and then, never having had the diligence to qualify for anything better, who'd been jailed.

Come to think of it, it must have been about a year now since I stopped seeing him shuffle drunkenly home after yet another day. I thought he'd found a wife. There'd been whispers about his involvement with some strange people but I hadn't paid attention to them. He was in some way connected with those goings on that seemed to be affecting everyone, even simple creatures like us.

These goings on had been keeping me awake for almost a full season now. I thought they'd end with the monsoon rains. But no; they picked up the season after and engulfed our world in a blaze of horror. I couldn't sleep for the disturbance.

Some nights, I'd be kept awake till dawn. Even the weeds growing on the sides of my banks dipped their leaves furtively and quivered unnaturally in the night breeze. I felt their disquiet, their uneasiness and agitation. Most times, of course, these weeds were a nuisance and I always told them so.

Very often I made it plain that I didn't want much to do with the likes of them. They were the insensitive, indifferent type.

But I saw that even they reacted to what happened as if what they felt was too much for them to bear. As if that were true, I muttered to myself, for I knew that they were common with no sensitivity whatsoever; only weeds.

But I noticed that they bore scars; their leaves were scorched and some lay battered and bruised as if deliberately stamped upon and then brutally pulled out and flung aside. They'd never suffered such abuse as far as I could remember. I was curious, so I decided to wait and watch and see, what it was that even run of the mill types like them, couldn't stand to see.

My heart knew it all of course. But I needed to see for myself, to convince myself that what was happening was for real. And I was concerned; as deeply concerned as some of you were wholly unconcerned.

So as I listened to Duh, I found that what she told me that afternoon was exactly as it was, exactly as I'd also found it to be.

One moonless night, I heard weedy susurrations, the rustle of anxious whispers as the weeds exclaimed in muffled undertones on seeing something that made them wilt with fear. I too looked up only to see, in a flash, what they saw: shadowy shapes of men cloaked in darkness and intent on murder. I was shot through with a similar fear, for I sensed calculated slaughters and cold intentions. And I saw, I saw what I didn't want to see.

The night was tattooing us with this ruthlessness; shredding us with this offensiveness.

Let me explain myself, our world is one of survival, of the big fish eating the smaller fish in order to survive, of the big wave smashing the small waves during a storm, of waters overflowing my sides and washing up river-food for small animals in a seasonal, natural kind of way, an acceptable state and fair enough for all of us.

But what I saw sent tremors, even for me, right down to my rocky bed. This was not survival in any sense of the word, the kind that we understand and respect; this was sheer murder. Even a snake preying on a chicken is understandable, for that's the law of nature –the bigger ones preying for food on the smaller ones.

This, what I saw, was none of this.

3

That afternoon, in my presence she let herself go; confused and confessing, lucid and rambling she poured her heart out. As she fell into a tirade again, I remembered events confirming her very words.

She spoke of the recent happenings; of covert whispers here and there from bystanders in the marketplaces, at funerals and alongside the narrow alleys that, like flowing streamlets slice and splinter your localities, where men would group together for small, nightly chats.

At the *jadoh* fast foods that thronged with people from all walks of life, there was so much apprehension that talk went on only in undertones. Every nook and corner of the *sor* was laced with fear. She would know; for after all, she scoured the streets of the *sor* looking for work, for herself and for her children. For her there'd never been any letting up in looking for any kind of employment anywhere. She was tireless. That was how she lived even then. And that was how she taught her sons to be. A wayward one here and there, amidst seven of them, however, was to be expected. Even she agreed to that.

There was hunger for change, she expressly emphasised that. She saw a rebellious anger, in some, who espoused change, change and more change. She quoted them; knowingly or unknowingly, accurately or inaccurately, I didn't really know, for who was I to question what she knew or what went on for that matter?

She never mentioned names but went on to confess (I pushed closer to her) that she felt responsible for pushing her son into taking up odd jobs with those who in the course of her search for employment, had inspired her so much. She'd been made to understand that they were her "liberators" who'd come to set the likes of her free.

Free from what? I wanted to ask. That's what she called them. I don't know why. I shall never know why. Perhaps you know why? But I was bewildered. And then she confounded me further; for she now complicated the world that I thought I knew so well, by splitting it into "liberators" and "oppressor dogs" and then splitting it further into something else that sounded like "the un-liberated." Did I understand her alright?

This woman was talking about things that I was only dimly aware of. What liberation was this? Did she even know what it meant? Surely, surely she'd been mindlessly influenced by such talk? And what new words was she using? Did she even understand?

She described the "new ways" as having burst upon them all of a sudden, overnight almost, gathering force as they swept along "lifting me, at first, with an enthralling sense of the new. But for the life of me, I never knew what they'd do to me. They swept over us like new brooms. I thought that they'd come to sweep out the filth of many decades and sweep in the new. That's what "they" promised me you know. They waved their words and their ideas at me and made me see what I never did! I forgot, forgot myself completely. I was swept off my feet. And I wanted more than anything that at least one of my sons would contribute to the birth of these new things, the shaping of these new ways."

Her tone had actually become wistful.

The poor woman.

Not only did she not understand the upshot of the new ways, I myself never quite understood them, but she'd risked her family's security in supporting and encouraging the one son who'd always been the laggard. Only now she knew that there could be nothing in it whatsoever; for her or for anyone else for that matter. Stupidity alright, especially in a woman like her, who had no one, no one at all to back her up! A single woman!

But was it stupidity for a woman to feel so passionately about something? Was it wrong to want her son to make something of his life after all?

She continued, "I didn't understand the implications involved. But when the boys began to disappear, my neighbour's sons for instance, snatched by the wind as it were, as if abducted by the evil spiri: *shah rah kla*, I was a little fearful, but didn't worry too much for I thought that these were natural processes in the shaping of the new. I was a fool don't you see? Can you imagine that? Can you see my foolishness?"

Duh Symper was questioning my ability to see! Little did she know how much I saw. Even littler did she know how much I knew!

But I forgave her, I forgave her, for I'd seen right into her soul and understood the difficulties that a mother like her had had to face with seven sons and a daughter to feed and bring up.

"In my ignorance", she rued, "I was hopeful. I hoped as I'd never hoped before for these changes to come, but even I was never quite sure what they were".

"And then one night, I noticed that my neighbours were having regular, nightly visitors who slinked in and out, unnoticed by many of us. They were thick as thieves, they who were like family to me and now who were so close-mouthed about it all, keeping everything away, even from me. Even me they never trusted!"

"So I eavesdropped almost every night on these neighbours and found that they were hotly debating many things that I'd never even heard till then".

"These faceless visitors, all men, seemed to be regulars. It bothered me for it was becoming too much of a nightly affair. And I was curious. I was also hurt that I'd been left out. And as I had an inkling of what they were about I wanted to show them that my son could be anybody's equal. He too could contribute to anything if he so wanted. I wanted my son to be part of this new planning, this new intrigue that went on behind closed doors."

She wiped a tear.

"It was easy to snoop on my neighbours. There could be never be any secrets in our barrack-style tenements with open corridors and washrooms behind. The wooden planks that partitioned us from each other had been put up a long, long time ago".

"They were rotten and would have fallen into pieces, had they not been propped up haphazardly by anything that was available to the many tenants, over the many years. Over the peep-holes that had been gouged out by earlier *lorni* tenants, pages of yellowed newspaper, sticking tape, film posters, plaster, anything available was used to seal them off".

"But one always heard everything that went on one's neighbour's side. I scooped up important information, learnt to train my ears to the inflections in their voices and heard how more and more people were beginning to co-operate with them".

I sighed with frustration for her and for myself. She was enacting a drama that she'd believed in, would have still believed in had her son played the game well.

"Who they were I never knew; they were as faceless as the devil himself, but I joined in wholeheartedly as I crouched forward enthusiastically, intently listening to their animated discussions from my side of the partition, ears glued to the covered peep-holes; my heart smouldering with imbibed passion."

What frankness! I couldn't help thinking. But I wondered how much of it she understood. Poor woman, I listened to her garbled recollection of those so-called discussions. How much of it did she really understand?

She never disclosed what the discussions were about, but I'm sure, that much of it went above her head. I knew that.

This woman who'd endeared herself to me by her frankness could also be cantankerous. Bringing up seven sons had done that to her. Friends were hard to come by, therefore, because the whole business of bringing up her children had taken away all she had. Perhaps that's why she was telling me all this. I didn't want to hear, didn't want to be party to any one-sided opinion, but as they say, how could I shut my ears?

How could I not open my heart to this lonely, single mother? For after all, would you have understood her better?

As it is, as Duh Symper confessed everything to me, I heard only the lies, saw only the pretentions that befuddled you and cheated the likes of poor ignorant Duh. She feared for her son. I found fear in my heart too, but I wasn't going to allow it in, no matter what.

Actually, what struck me as I listened to her was that I too had been made a fool of, like her, just like her. For as long as I'd been in existence, I always knew that you'd kill even for petty reasons; but I never expected you to use me in the way that you did now.

I wasn't even discarded or rejected. I was simply covered in your refuse; overwhelmed by it; up to my neck in it, chock-full of steamy leftovers. I found I could breathe only when the rains came. But even they refuse to come as they used to; birds don't fly anymore, shot down by such as you, there are no fish now.

And Duh here was telling me things that no one had told me before. Duh and I, who were we, simply two creatures used and discarded by the whole lot of you. I felt as betrayed as she, as vengeful as her.

April showers and I! I never imagined we'd make such a compatible twosome. Did I too want to be "liberated"? I questioned myself, as I half-fearfully introspected her words.

But there it was. I was as cornered as she was.

4

"Bahdeng Mawlut was a trusted friend", she continued, as if she were following a train of thoughts that she expected me to follow instinctively. Presumptuous of her! I tried to keep up with her, tried to tie up all the loose ends of her story but I sensed a sudden and swift change in tone when she mentioned Bahdeng Mawlut's name. Was it fear, was it awe or was it disdain? So, she was afraid of someone! And she was revealing herself to me bit by bit.

"He was my Nang's friend. I trusted him". The story was coming out once more! This Bahdeng Mawlut had affected her in some way, I knew.

"Poor folks like us you know, are closer to friends than to relatives, especially when all the relatives that you have, are rich and snooty and pretend to be superior low-born ingrates all. They come from the same stock." Here were the tobacco-laced April showers again.

"Because we're in and out of each other's houses all the time, we feel more comfortable with each other"

"His *daju* father worked hard in Ïewduh", she carried on.

Ah! I remembered him! I knew that he navigated its alleys with tons of stuff piled high on gunny bags shoving and pushing; fighting for space even as he hauled his heavy loads.

"His mother was a *nongdie mar sekon han,* vendor of second hand goods", she filled in the details not knowing that I knew more than she ever would.

She roamed the localities of the city, with a *khoh* on her back, scrounging for old stuff, buying from her rich clients who hoarded the used stuff to sell to her, so that she could sell it back at Ïewduh.

"Come to think of it", she continued quite honestly and yet shamelessly too, "I was always rude to her even though we carried water from the same tap and used the same latrine. Hmphf! But sadly she doesn't even have a daughter" she concluded on a compassionate note.

As for me, I knew them well even without Duh's detailed introductions. But why was she relating all this to me? And I couldn't stop listening even then. Maybe I really am a *lorni*! But what's with that? Listening to others never harmed anyone, so I continued to listen for I found that I was gradually agreeing with her on more occasions than one.

I was never ever at ease that afternoon, but I did find connection with her – in her aloneness, which I resisted. I didn't want to identify with her all the way now. I fought her with all my being but I had to listen, as always. She'd already dragged me in, this way that way.

Anyhow, Bahdeng Mawlut's folks whom she was describing in such detail were people I knew like the tail-end of my waves. They presumed that I was there only for them, only to serve their requirements. *Shish eh*, they used me as if I were the dumping ground itself. They only saw me as a useful resource to be used unsparingly and uncaringly.

Somehow they thought they'd find shelter beside me. I knew they associated me with the things that they'd left behind in their village. But, they never gave me their hearts, never entrusted their souls to me. They were insensitive country migrants that I knew only too well. But Duh didn't know that.

And my banks were dotted with their tenements. I prefer not to tell you the name that they gave their locality.

She spoke of the many sons whom many mothers had been robbed of, "the Bahrits and Bahdengs and Bahduhs and Bahheps lured away from their families. And as soon they were lured away they did a sudden crossing over that changed them almost immediately".

"It was as if they were given new bodies, new minds, becoming new entities who were ready to take on those whom they were told, opposed this newly emerging community of real men". She was silent about her son, however. I merely waited for her to continue.

"What were these new ways?" she quizzed herself, almost as if it were a game she was playing with herself. But it wasn't a game. She knew it. I knew it. And yet "it was as incorrigible as the see-saw game that I'd often watch children play on their school playgrounds. One minute up, one minute down; victory defeat; trust betrayal; up down".

Partner-predator, I wanted to add.

"See-saw".

I too saw it all when she put it that way. And I also saw the rifts that divided and further sub-divided and caused fissures in all your lives. Nothing had changed really, only this time the dividing fissures sub-divided again so fast, it was impossible to keep up with them.

But what did she see and hear? Many things that were now clear as day and audible to her – through her son's friends and acquaintances, through their seeming involvement with something far more important than any one of them, individually. They evoked her respect in a way that no one had ever been able to.

She was a dreamer, this one, like me! So she continued to put more pressure upon her son. He was the only one who listened to her completely; Nang the laggard who was too lazy to bother with things that required him to make an effort. So he allowed his mother to nudge him bit by bit to do what she said was good for him. At first it was only to stop her nagging him, then it was just for the drinks.

I was listening to a mother's confession and I had no right to find fault with her. She needed another woman, another mother to listen to her and as I filled in as one for the moment, I began to understand her side of the story.

5

And then I found that what she told me was something that I'd already known.

"Bahdeng Mawlut had two older brothers who'd been born one after the other; the result of their parents' ignorance and youth. When they were older, they'd run away from school to indulge in their usual pastime of running off from everyone and hiding somewhere".

It seemed that early in life they'd learnt to shirk responsibility, I added, more to myself than to her. They'd often hide in a shallow cave hidden somewhere in my recesses. They found it one morning when they were still waddling babes, when they were playing all by themselves, as they wandered off alone; they were used to being alone, used to being neglected by parents who wanted a daughter so badly, whose hands were always tied with other responsibilities.

If I'd grown sluggish with filth that wasn't their concern, they were too young. But even when they were older, they were indifferent to me, indifferent to everything about me except that I gave them easy refuge.

Their parents didn't pay much attention to them, so they were left to themselves all the time. I'd listen to their childish prattle and watch the unending games they played all day long. Even as young boys, when their faces had no facial hair whatsoever, whenever they got money they'd run off to buy *kyiad*, hide here and drink their bottle in imitation of the only men they'd seen in their young, lives, until they couldn't stand up.

If they did, for sure I'd drown them. I'd see to that, I often promised myself. I couldn't stand to see what I saw.

She conveniently omitted the fact that her Nang very often drank and had fun together with them. I'd seen him, so I knew. He was their friend. But he was also their bother, Bahdeng Mawlut's friend. I found that I was filling in a few gaps in her story myself.

So what was she trying to tell me? What was she trying to tell herself? There was an opaqueness there that wasn't quite her, that wasn't normal for her. She was hiding something from me, hiding it even from her own self, shielding herself from some inner-knowledge. But being Duh, it'll soon see the light of day. I knew that. But I'd have to listen more carefully now for I smelt something fishy, sensed an uncanny something that disturbed her even more today.

"Bahdeng Mawlut", continued Duh, "was the only one who seemed destined to fulfill his ambition. He was determined to make something of his life".

It was getting cold. Duh shivered involuntarily as she squatted down behind a bush for a surreptitious pee down my banks. She took her time. And I waited as I always did till she stood up again to make herself ready with the next mouthful of telling.

As she was settling herself down once again, for she was wanting to tell me more, I too recalled Bahdeng Mawlut, the youth. But at this time, more and more, other concerns were also occupying his mind.

He was at a critical time in his life, vulnerable to many kinds of influences. As a matter of fact there were many others like him, half-grown, partial men who'd meet every evening after supper, to chat and talk and share the evening with each other. They were equally vulnerable. So why didn't she talk about them, mention them by name as she did Bahdeng Mawlut?

They were young; some idealistic, some rowdy some untaught, filled with the untamed energy of youth. Most of them were there because their homes spilled over with too many siblings. Their only escape from the overcrowdedness was to meet like this every evening, just to talk and pass the evening, smoking *biris* if they had the money.

Duh slipped in the information that her Nang was always part of this group.

For my part, however, I remember relaxing with them, yearning to ruffle their hair to assure them that life wasn't bad at all, happy that they were still simple and good-hearted, happy in the knowledge that this peace would always be a part of us in these hills.

"Bahdeng Mawlut was the only one who struggled valiantly to stay on in school. The rest had given up completely, their families needed the money that they could bring in and they were employed here and there as casual wage earners in the city", she said.

"They met as they always did to chat and talk and swap stories and jokes, and share in a togetherness that radiated waves of warmth and security especially on cold winter evenings around a small fire".

I knew what they didn't have, I felt like telling her; I knew it all and I understood their need for this kind of camaraderie.

And I would sigh with happiness to have them there beside me, I just couldn't help it; they were good for me. I preferred not to think of their future. The present was enough for me.

They're so ignorant and young I kept thinking, and so vulnerable. These were boys who'd grown up with each other and meeting up together was as natural and instinctive as breathing itself. There'd been other boys before them. There'd be others after them too.

I delighted in the company of these young men who, to me, was the entire world, who belonged to my world. I heard their easy laughter, shared in a spontaneity that made me young again and approved of their readiness to share each other's problems. I wanted to stop the seasons, stop the flow of my currents so that I could have them there with me forever.

I never knew that this would be lost, so soon.

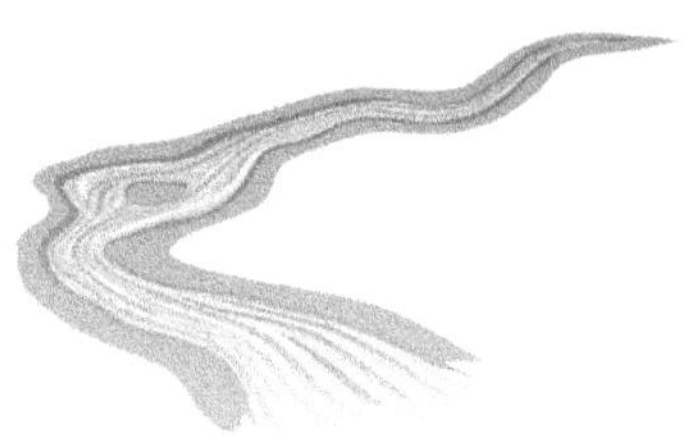

Chapter Nine

Duh Symper and I took over from each other at sporadic intervals of telling. Her stories joined mine intermittently; hers flowing into mine, mine into hers; mingling with each other till they met to flow on, in one thick current of intertwining thoughts and sentences that sometimes struggled to outdo the other or to complement one another.

So it was, that I now felt a certain solidarity with this woman whose tongue could be as caustic as the *shun* that was used to destroy the parasitic moss that stuck to the concrete in the rainy season; as bitter as the tobacco that could dislodge the most stubborn leech that clung to your calf but as occasionally loving as a mother's should always be.

Her stories sometimes grated with a sourness that clung to her soul. I felt it scour my own story-line. But I let it be for after all, her life had been rough and she'd had to learn to tackle it bit by excruciating bit.

"All of them, all the boys were conquered by this stranger who'd apparently befriended them; who", in her words, seemed to have emerged dramatically from the darkness "like one of those undefined water spirits that haunt some coves".

She described his voice as being "low-keyed, pulsating with some unfamiliar energy, deliberately persuasive".

Where he came from she never spelt it out to me. She was too agitated. But I knew of course nothing could be as dramatic as she put it. These incidents had started intermittently as I told you, here and there and had gathered momentum over the years.

"Those nights of passionate conviction, out in the dark cold, those were nights I've never been able to forget".

She spoke of those nights almost as if they were nights of initiation. Strange woman; I'd caught her before, now I caught her again. She was re-enacting her story as if she didn't want to let go. It was as if she craved to show me her part in it!

It was becoming the story of *her* life. Yet I never saw her on those nights. I was also there you know. Had she been hiding and eavesdropping even then?

I for one have felt that what I saw, what I heard those nights, were as unreal as the *thapbalong* that blurred the night with their mischief. Did Duh hear it all? I'm sure she didn't!

I too had heard him talk of things that these half-grown boys, partial men had never heard before. I'm sure he must have been one of those regulars in her neighbour's tenement, passionate and articulate; the one whom she must have often heard; else how could she speak so familiarly of him?

He must have had instructions to speak about such things, using the ways and manners that he did, to entice followers and listeners, such as Duh Symper all too willing to eavesdrop on her neighbours all the time!

He measured every word with stringent care. I too was fascinated. I too was captivated. I too was unwillingly drawn into his little circle; mesmerized by his open talk of change. I knew, I knew alright what he meant by it all.

But the boys didn't; they were simply won over by the magic of his words. He wanted them on his side; he would have them on his side. I knew that, as much as I knew that I too was hooked to his words. They were being fed a different fare. They needed to digest it even as I digested the words of a stranger who seemed overly confident.

I felt a mix of new emotions that sent waves of exhilaration right down through me.

Duh continued to confess to me that she'd listen to him through the rotting partition in her tenement, from time to time at first, but more and more often, as she found herself hypnotised by what he said. She never saw him. But she'd learnt to recognize his voice. She was as much taken up by him as I was.

He'd taken complete control of us! We were no different from the boys. An unlikely, infatuated twosome again!

And that was that. That proved to be the end of the old ways for many of them.

"He dogged them and stalled them with his cunning before hijacking them, even the conscientious Bahdeng Mawlut, with his words and his money into the new ways" she conclusively stated.

That Bahdeng again! So often did she mention him that I heard alarm signals ringing everywhere. He was like a spoke cutting into her.

Where was their native wit? I asked myself, not realising that they'd lost it somewhere in the stranger's strange words, as we already had. I felt the goose-bumps come, knowing for sure that this would be the beginning of the end, not the start of new beginnings as they were made to think.

The embers in their hearts had been stoked and a fire had been lighted. I felt its shimmering heat rise to meet the darkness around.

"It drew strength from those Bahrits and Heprits, Bahdengs and Bahduhs who were promised things they could never imagine nor ever hope to imagine, in return for their youth and passion, their time and enthusiasm and if I may say so their souls!" further added Duh.

And truly their dreams came true at first; as ours, Duh's and mine, also did; at first.

2

"That was about the time of the disappearance of young men, snatched even from under their parents' nose. I didn't understand much except that old Kmie U Hep's tea-shop began to buzz with covert talks of thwarted assassinations and other unwonted activities. People came home early in the evenings, for the dark was wont to throw up all kinds of strangers up to all kinds of activities. And the night was no longer what it used to be" rushed on Duh.

Days seemed normal; there was the usual sound of laughter and young people falling in love. But something was amiss. I smelt it in their breaths, felt it in their racing pulses. It breathed hoarsely down upon me. Conversations were toned down. The carefree ways had disappeared.

When night came there was an eerie silence: no shouting, no easy teasing of friends, no lovers taking a last round before going home. Sometimes there was that lone drunk singing to himself, lost to the world, but one would never hear of him again. What had happened?

"Late at night cars would be about their business. Strange business I observed thoughtfully to myself to be out so late, the sound of engines penetrating the deep night. And I also noticed that when these cars were heard in the distance, people hurriedly switched off their lights as if to perpetuate constant darkness. Why didn't I take note then, why?" she questioned herself frustratedly as if to turn the seasons back for her son, only for her son.

I sighed with despair for I empathised with this woman who loved her children more than anything else in the world. I too have this capacity for love.

Duh was awash in her emotions. She sat immobile, transfixed and locked in the past, strangely vulnerable. I felt a rush of sympathy again.

3

But let me add this at least, to make the picture clearer to you, that here were panthers on the prowl, hungry for prey. And the victim could be anyone whom the panthers sniffed out. They were as hungry and as cunning as they were sure-footed.

I saw the motiveless violence that their hunger brought on that caused me to inadvertently swallow spilt blood, making me gag with the taste of it. It seemed to me that there were many more panthers than there were victims. And I was forced, forced I say, forced to witness repeated bouts of senseless violence with a certain pattern for sure: partner-friend one day, panther-prey another; partner-friend one week, panther-prey another so that the whole world seemed given over to this strange kind of relationship – unforgiving and revengeful. It was betrayal of the cruellest kind: of panthers honing in for scapegoats; disguised in darkness even in the day.

After nights of bloodletting, I'd see men in similar looking clothes, uniforms they called them, swarming the undergrowth like maggots; crawling in my vicinity. Perhaps they searched for dead bodies to feast upon? Were they panthers or victims? Frankly I could never make out. Perhaps they searched in other places too where I couldn't see? But for whom? I never knew. I can only tell you what I saw around me.

They were probably searching for evidence of those panther strikes – that, I got to know as I watched them closely. They never found any proof whatsoever, no pugmarks anywhere in the weedy jungle. It was always like this. Even if they did see they pretended not to see. I was dumbfounded.

Of what use were their eyes then? Were they abettors in the bloodletting? The game was getting out of hand it seemed to me! I felt an uncontrollable shiver run up my spine as I uncharacteristically smelt my own fear. I was allowing myself to become a victim even like those others whom I pitied so much. But who would pity me? I'd pondered over this many times before and came up with nothing.

Nothing at all. No one at all to pity me.

I felt the fear that riddled Duh. But for her it was increased tenfold because she had her sons to be afraid for!

But, I spiritedly told myself, I had to be strong to deal with preying beasts such as these. The only way was to tread cautiously, as cautiously as the prey themselves but without their fear. Can I? I asked myself, I'm not their prey. They can't hunt me down. So I reckoned hard, they'll have to deal with me on my own terms then.

Meanwhile, I told myself, I'll simply wait and watch. But if I find wrong moves, wrong actions anywhere, I'll do what I did to Khrawbok. I felt better for having confronted them, their ruthlessness in my mind.

On this matter, Duh I'm sure was not as fearless as I was.

They would describe it as a hit and run kind of violence. But I could also see that they were also afraid of finding too much, afraid for their own lives, afraid of shadows even when the sun was high. There were panthers there too who'd infiltrated their ranks and camouflaged themselves in the same uniforms. This game of hide and seek had to continue for the sake of those who desired that the game be played, who pulled the strings and seemed to don the same clothes; who planted stooges in uniforms.

I was getting to know the game so well I couldn't help wondering how some of you who played it with practiced adeptness, played it so dispassionately indeed.

As for me, I could anticipate your every move.

The rest of you, I pitied. With us folks at least there was a simple law to be followed. Not so with you.

This was some bizarre game that didn't follow any rules at all. What were rules anyway? Just terms and conditions to show off your superiority, indulging in your favourite pastimes; games after all! And you make such a show of actually following these rules!

But, mind you I know! The rules that you make are shifting and temporary. The moment the games stop serving your needs, you stamp them out, stamp them spitefully, deliberately; as arbitrary as that! What happens then? *Phuit eh*! Games without rules indeed!

I could see the cracks coming on even more and the snaking fissures dividing and sub-dividing; dividing and sub-dividing further.

4

It was later in the evening that Duh began telling me how she began to discover many unpredictable sides to the boys she regarded almost as her own, "the ones who'd grown up with my sons".

She spoke of Bahrit, the same Bahrit whose hair I'd want to ruffle several times before on those cold, vivid nights, nights of friendship and warmth. He was funny and full of life and always up to playing pranks on others. He never meant any harm at all.

But, "he'd been overhauled completely. His body was now taut with discipline and training. He was handsome; trim and muscular, the result of endless exercises in forest camps and strange hideaways. There was arrogance in him, however, newly put on, that was startling to a person like me who'd known him so well; that surrounded him like a brittle shield."

She couldn't possibly know that with great regret, I could have also added much more to that. But I'll say only this, that I'd seen him make a come-back after a mysteriously long absence only to rule everyone with "the look". This was the look that distinguished some of them from the rest of them.

Duh too, saw not only Bahrit, but Bahduh as well change to an unnatural degree and swell up with an unnatural kind of pride. They seemed so sure as if they'd accomplish something and were about to accomplish something more. No more vacumn in their lives! They carried themselves confidently, injecting fear into the ones they'd soon single out as the victims.

"What accomplishments?" spat out Duh venomously for she scorned their involvement in matters that, she said stank with perverseness. She must have known what she was talking about.

I must say, they were diligent and skilful, these two, sharpening their claws and honing their sight to ferret the mice from their holes like cats at play, before pouncing on them. After each exciting game they'd lick themselves clean and stretch out in the sun, tired out with the dark challenge of their sport. Their bodies were filling out and soon they'd be full-grown panthers.

Already their names cast fear and they'd begun to scout for the prize-winning prey hidden in the jungle that they'd help create.

5

"Heprit of the red *tapmoh*", spat out Duh, speaking with mixed emotions again; her loyalties now seemed confused, "was as indiscreet as a howling dog at night, a hungry snorting pig, indecently publicising his importance now".

"He was one of many who were consumed by an exaggerated sense of self-importance. These types lost their usefulness in no time at all. Heprit of the red *tapmoh* was brutally assassinated by his own friend, Bahdeng Mawlut who killed him out of a sense of duty".

"They have wheels within wheels' she whispered.

I whispered back, is this implicit justification for the assassinations?

6

"There'd been panther strikes again" said Duh.

Yes, I too recalled them.

But wait, I never knew that she could hear me.

She actually referred to the panther strikes! I was elated. She'd heard me!

We were on the same wavelength then, this lonely old woman and I, lonely ancient river, who saw things that no one else saw.

As she recalled it, I too remembered the buzz of covert whispers criss-crossing me as if the world itself was on fire. When I strained to hear something, I heard an indeterminate buzz and saw only subdued activity. Some things were going on that I needed to know, that turned life upside-down.

Duh was deeply troubled beyond my understanding. In her impassioned recollections, she seemed suddenly to have moved far away from me. As she spoke of the men in uniforms finding a body abandoned in a ditch far away from my reaches, she described the murder in such detail that I found it impossible to imagine anything like that ever having taken place.

It had affected her to a terrible degree and as I understood from her, the gaming-wheels were clicking fast, clockwise, anti-clockwise; clockwise, anti-clockwise, taking down victims right and left.

"Who was to know that it was the body of Bahdeng Mawlut except for the murderers themselves?" she asked herself agitatedly.

She was given to letting out details bit by bit, as if held ransom by it.

I smelt a something there, so I waited for her to go on. Her obsession with the murder was inordinately exaggerated, and yet . . . . she couldn't seem to snap out of it.

"Bahdeng Mawlut's parents were totally ignorant of the panther-games that their son was into". That, I'd known for sure.

"Their shock was terrible to see. They were simple. They had their faults. They'd committed their own errors. But life wasn't so simple".

Didn't they with their long experience know? I'd wanted to ask them this myself.

"Bahdeng's funeral" she spluttered incoherently as she grew more agitated, "was to be held the next morning for the body had decomposed to such a degree that it was impossible to keep it for two nights, as is the custom. I was unable to weep, for I'd seen how he'd been transformed."

I knew what she was trying to tell. He'd become the largest panther with the most trophies and the largest lair.

"He was to be finally put to the ground as he himself had done to so many", she continued strongly. "He'd put away too many in such a short while. I'd known him well but never anticipated this falling off,"

into prowling predator , I wanted to interrupt,

"where even I was wont to fear him with an irrational fear. I felt avenged, yet curiously unable to hate him now". This woman was contradicting herself! One moment she hated him, now she pitied him. What was this? Surely she'd been scorned? How had she loved him then? As a son surely; *hooid*, yes, it dawned on me; just as a son; for he had been her Nang's friend.

And things became clearer to me as I listened and sympathised and empathised and added my own pieces to the puzzle.

It was our puzzle now to complete.

"I noticed with awe that the coffin was made of *kseh saw*, seasoned red timber that was carved and polished almost to perfection. He lay in state like a king to whom minions would come to pay their last respects! But where were they?" questioned Duh.

She showered me with frothy spit as she passionately re-enacted the drama that accompanied his death. Duh here, seemed to be saying something about Bahdeng's death. Did she know more than what she confessed that evening, which was why she was so deeply disturbed?

"I waited the length of the funeral for them, but no one seemed to so much as even want to come close to it. It was draped by a black flag" she said.

"How it came there nobody knew. It seemed to me that no one dared say anything outright about the unexpected things that seemed to be happening at the funeral, that day".

Many panthers were at large. They posted themselves strategically amongst the crowd and tried to mingle with them. But fearing-to-be-victims-people sensed them and shifted uneasily wherever they were, snorting nervously, ready to sprint off any moment.

This she didn't know because as I keep harping on this one point, she was limited in her seeing. I was not.

"Some came to pay their last respects perfunctorily or simply to *lorni*, to gossip and tell others later. There was a small handful, however, that actually came to mourn for him. There were well-wishers and hangers on, gossips and onlookers, but they definitely stood apart"

from the panthers I silently added.

"in the innocent way they carried themselves and in the easy and friendly manner with which they talked to one another".

But only I, with my river-sensibilities, felt that palpable sense of relief in those who were there, a certain lightening of the heart as it were. Were they glad that he'd died? Then why pretend at all? Why all this sham? There was also an air of anticipation, I'll never forget it, as of, a waiting for something to come to pass.

The panthers I instinctively knew. A whiff told me where they'd silently posted themselves; Bahdeng Mawlut's panther-partners. Was he aware that the equations had shifted from partner-friend to panther-prey? Which was why he was murdered? That they stalked him down even after all that he'd accomplished for the cause of their future lives?

Duh tried to speak of many things, but choked with too much emotion. Maybe she understood only too well what the panthers were now up to, having been an intimate but liminal part of their nightly discussions? Maybe she was worried sick for her Nang?

But the ones who once used me as their refuge and dumping victim, Bahdeng Mawlut's two older brothers – they never attended.

7

I took over from Duh to speak of things that affect me even now. She'll have her last say, don't you worry. She'll never be at a loss for words. She stakes her own claim on all the speaking words that her tongue can appropriate! I'll allow her that. That's why so many have to think twice before coming up with a rejoinder. It'd be well to learn a few skills from her you there struggling on alone in this world!

Even a week after the funeral, I couldn't tear myself away from the family. I didn't understand why. Friends and family continued to keep them company over those dark, dismal days. There was secret, secret talk about the strangeness of his death. No one seemed to be able to put it away.

Bahdeng Mawlut's parents' house was milling with neighbours and friends and relatives who'd stopped visiting them all these years. This was chilling and strange, for I knew that almost everyone had avoided visiting them for a long, long time, for reasons everyone kept to themselves.

Why were they defying their fears and coming to Bahdeng's parents' tenement again and again even after the funeral? It was high time they went back to work! Yet they stayed on unable to leave.

And I discerned talk, low talk of the killings and murders that had taken root in this soil, our soil. And something else caught my attention. My ears pricked up.

Was it fratricide?

Did they say fratricide?

I breathed in sharply, horrified by it all. I hovered closer to a group talking in hushed whispers, bunched up outside the house cutting *kwai* nut and dressing the *tympew* leaf with the caustic white *shun*. They'd been part of the community wake that kept vigil with the bereaved family, but they seemed not to have been able to pick up the threads of normal life as yet. They were glued to where they were a week ago. But what I heard made me retch with a mix of emotions I'd never felt before

"But the truth will out". I heard it said by all of them.

As for Duh and me, we knew; instinctively.

8

One night I heard those two again.

Bahdeng Mawlut's two older brothers had come back to me. They sought the shallow cave that was once their refuge when they were waddling babes. They'd come home after all. They needed to hide their grief. I felt their uneasiness, their fear but most of all their despair. I was also mesmerised by their actions. They were wont to lose themselves in drunkenness and bitter accusations. They seemed only to talk about their dead brother: about how they never recognised him, how they were blindly led to it, how they wanted only to hide, hide, hide. This time the hiding was for real. This was no childhood game.

And then I pieced it out bit by bit. And I found that I was afraid, afraid that what I'd heard was true, afraid for the likes of all the Bahduhs and Bahheps and Bahlits; afraid for all the victims; afraid for myself; afraid as never before for what awaited Duh's son.

Afraid for us two. Mindlessly afraid.

9

I also knew that the panther-partisans who'd commissioned the two for the murder were hot on their trail. And they were close. They scorned the killer-brothers for their weakness, for having babbled about it and made sniveling idiots of themselves.

Their fratricide had stricken them with remorse, fried their insides, turned them inside out and rendered them useless, reducing them to a state of severe despondency.

"Anyhow they've outlived their usefulness. Besides," growled the king panthers to one another, "we were considerate enough not to let them know whom they were assassinating. If they recognised him in the act of disembowelment it's nobody's fault but their own."

The hunt for them was brief but intensive.

They were eliminated in the cave, my cave; and their bodies were left for the four-footed predators after all.

10

Duh had known once and for all, that afternoon, that the eagle and the crow could no longer be silenced. She was crushed completely.

Broken as she was, I realised then that she'd left many things unsaid, unable to speak about things that were too close to her heart, things that troubled her very much. As I waited, I saw how shrivelled and small and dry she'd become.

I could have swallowed her whole, on a thunderous, monsoon moment when my swollen belly belched out with the pounding currents of unrelentless rain. But as I waited and listened that afternoon, I felt that we'd become so close, she revealing all her fears to me that we were like two sisters, two creatures tied together by a common understanding of the world!

So much for maturity then, if this is what it does to you – make you dearly feel the cost of life, the cost of friendship, the cost of understanding.

It would soon be dark and I knew that Duh would have to leave. She was restless inside. I knew it. I felt it. She was unwilling to go, unwilling to let go of me, unwilling to leave this temporary sanctuary that she found with me. For where would she go?

None of her sons had ever been able to hold out any comfort for her in *that* way; she'd always relied on herself, relied only on her own resources. They were only her sons after all. She never expected too much from them. She always allowed them to know that she was self-sufficient. It would never occur to them that she needed them *this* way.

As for her daughter, she was in the thick of all kinds of family-responsibilities with her children still small and her husband struggling to make ends meet. Poor helpless woman, I thought dejectedly.

Frustration and helplessness seemed to pile onto her, almost drowning her completely. She was pushed further into herself as she realised that her adversary was not in the gossip that hounded her these day, not in the isolation that surrounded her, not in the circumstances that she now found herself in.

Her adversary took the shape of her son; the one son who'd been too lazy to think; the one son stupid enough to mess up his role in a game that he only feebly understood.

And then I saw the furrowed brow that gashed her outwardly and gored her inwardly, despite the bristle in her voice, despite the show of strength, despite the undefeated attempt to challenge the powers that be, I knew then, what it cost to be a mother; especially for someone like her.

Nang had made it a habit to mix more freely with everyone. He associated himself with people he knew well and with people he didn't; making it second nature to him as he penetrated every social nook and cranny. Was he following someone's instructions to the last word? Or had he become gregarious overnight? His actions seemed premeditated; uncharacteristic for someone who wasn't calculating by nature, who wasn't much of a talker or a doer either; and yet and yet . . . . Had he finally been pushed into playing intricate, dangerous, abetting games?

Or was it, "anything for a free drink" all over again?

Duh Symper knew that her Nang's incarceration at the District Jail was linked to the murders of Bahdeng Mawlut and his two older brothers. She had neither riches nor powerful connections in the kachari to buy her son his freedom. She was beaten alright but was stubbornly holding out against the world, all by herself.

Bit by bit, life had whittled her down by slow degrees. None of the other sons knew what to do. They never had a mother's instinct after all. They'd heard, of course they'd heard and they'd come when they heard, but they too were stumped.

Now it was only Duh and me who were left together of course; Duh struggling to master herself, me struggling to make meaning of it all.

It was the sense of a game again. And yet it wasn't. "See-saw" as she once said. So, "see-saw" it was for Duh. When Nang was "conscripted", as I'd heard the word used by all of you, she was overwhelmingly up. Now she was humiliatingly down.

Her Nang's hands were tainted, steeped in the blood of the friends that he'd grown up with. He'd led the murderers on with information, indiscriminately, for a price.

Who told her all this? None other than the eagle and the crow, that brought to her information that others already knew. A few friends and well-wishers tried to keep it from her in a bid to protect her! But she inevitably found out and see-sawed between shame and horror, pride and wretchedness, up-down, see-saw.

Her son had turned out to be a treacherous-double-crossing, treacherous-informer; who'd traded "anything for a free drink" after all. She just couldn't hide the fact any longer. And it broke her as nothing else could. What could I offer her? Only my own feelings, as broken as hers.

I felt her crumble; her weaknesses becoming more obvious and her feelings that evening, more transparently fearful.

She found that she was back to where she was, when she'd started out several years ago promising herself that she'd be both mother and father to her children. That seemed like *mynhyndai bah,* aeons back when she was all alone; only with her children and no one else to protect them. Now she felt she was back there, where she'd started on her long journey. But this was not the starting point, this was the end point.

"Phuit!" she spat out, nevertheless. And then, a shower of words as she raised her head again, unbeatably:

"I'm going to take them on single-handedly, the whole kachari of them, all lawyers and police and all of them. Let's see about that! Let's see them try anything on my son! The salt of my life, the beacon of my soul, the one son who's done me proud! Even as I stand here this very evening, I dare anyone to challenge my right as a mother; I speak for my son. . . ."

Like a cornered animal making a last-ditch stand.

But now I knew her, I'd fight with her. Only her love for her children mattered. Only her loyalties mattered. Nothing else mattered.

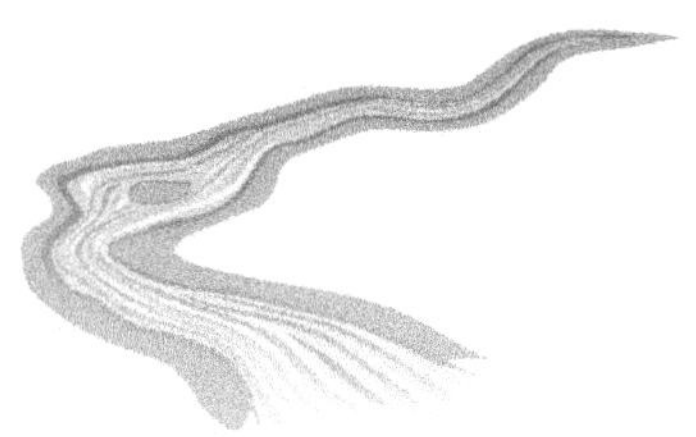

EPILOGUE

So what else do I tell you? What else do you want from me?

After all I've seen, after everything I've been through, after what you've implicated me in, games and the shadow of those games, your shadow casts darkness upon me.

I'm overlaid don't you see by the in-dissolvability of your crimes. They weigh me down as if to drown me out.

But I refuse to be put down. I refuse to be rendered useless. I refuse your illusions. I all.

You've used me to prop yourselves up, to justify yourselves, to flush your crimes down my belly. And even as you stifle my breath, my very soul, you must know that I'm destined to live as your struggle to live is inextricable from me.

Yet I'm still haunted by what you could have been, mesmerised by the visions that you create and destroy, the dreams that crest your fallen rainbows.

Can there to be no end to this, no end to the wretchedness that you inflict upon yourselves with your hankering after that which you selfishly call life?

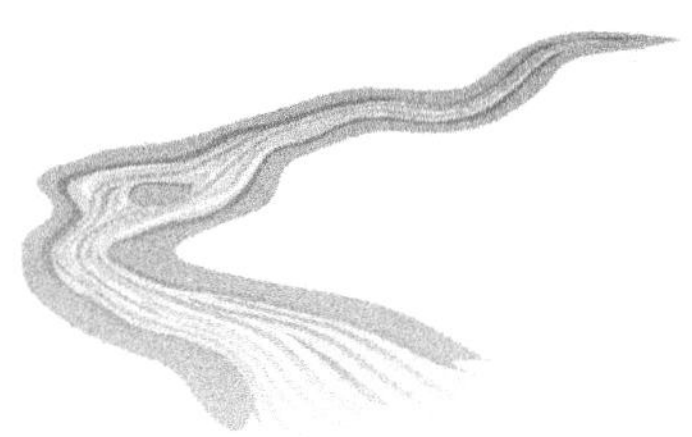

Glossary

bapli	poor thing
bihari misteri	bihari carpenter/mason
dabor	basin
dhara	Khasi formal wear originally consisting of a 6 metre silk cloth draped around and fastened on both shoulders. Nowadays two pieces of fabric are used.
dieng	trees
dkhar	person from the plains
dohjem	pork or beef entrails that are a delicacy
dohlun	tadpole
dohsnier	pork or beef intestines that are savoured as a delicacy
dohtyrkhong	smoked meat
duli	small almira used in kitchen
em	isn't it? no?
ieit klim	adulterous love
Ïew Polo	Polo Market (*Ïew*)
Ïewduh	name of the main marketplace in Shillong
ja sngi	midday rice/lunch
jadoh	pulao
jaiñkyrshah	Khasi apron tied in a head on left shoulder
jaiñsem	Khasi formal wear like the *dhara* but made of material other than silk with no tassles at the bottom
jaiñsop	Swaddesy blankets/clothes
jingdukhi	obstinacy
kait syiem	a type of banana
kha mukur	magur fish
khasaw	trout
khasi Mama	khasi uncle/ big dad
khaw-ot	undercooked rice
khoh	khasi basket carried on the back
kseh saw	red-pine
kwai	betel-nut
kyiad	local Khasi liquor/ commercial liquor
hooid	yes

lakait	banan leaves
ka 'tiew lalyngngi pepshad	refers to the legend of the flower that spent all her time dressing up and sprucing herself until she missed the dance; refers to a procrastinator
lorni	nosey-parker
maw	stone
mei-i-thei	mother of eldest daughter
mynhyndai	ages back
nepali daju	nepali porter
ñiang mong	pits that form on the face of a woman especially during pregnancy and childbirth
nongdie mar sekon han	vendor of second hand stuff
nongkyndong	village/villager
nonqsor	City folk
nongtrei kynta	cleaner who gets paid hourly
paitpuraw	the bulbul
pata kyiad	place where *kyiad* is sold
pla Ïew	cloth bag
puri	river nymph
rang khadar lama	a womaniser
shah rah kla	refers to a prowling spirit who is supposed to kidnap a guileless person especially on foggy nights
shalynnai	tiny fresh water fish
shersyngkai	another variety of fresh water fish; not as mobile as shalynnai
shun	type of lime used for scouring the moss that builds up on cement during the rains. The other type is eaten with *kwai* and *tympew*
sohkheiwja	bits of cooked rice fallen on the floor
sor	town
suka	25 paisa
synrei	pestle
synsar	broom
tapmoh	Khasi shawl
teem	teer gambling
thapbalong	devilish phantom
thlong	mortar
thwei	river depths
tungrymbai	pungent Khasi dish made of fermented soya bean
tung-tap	dry fish
tympew	betel leaf
tyndong sliew-ding	short, cylinder/bamboo/metal used for blowing on fire
tyndong tympew	bamboo cylinder where betel leaf is kept
tyngier	bamboo frame over the fireplace
u pyrthat rkhiang	dry thunder
umthlong	whirlpool
wah	river

www.ingramcontent.com/pod-product-compliance
Ingram Content Group UK Ltd.
Pitfield, Milton Keynes, MK11 3LW, UK
UKHW021011290726
14059UKWH00001BA/77